I0687973

James Macdonald Oxley

Up among the ice-floes

ISBN/EAN: 9783337107918

Printed in Europe, USA, Canada, Australia, Japan

Cover: Foto ©berggeist007 / pixelio.de

More available books at **www.hansebooks.com**

Up Among The Ice-Floes

BY

J. MACDONALD OXLEY

Author of " Diamond Rock; or, On the Right Track,"
&c.   &c.

T. NELSON AND SONS

London, Edinburgh, and New York

1894

CONTENTS.

UP AMONG THE ICE-FLOES.

<hr>

CHAPTER I.

ABOUT TWO IDEAS.

CAPTAIN MARLING had an idea. So too had a sturdy boy, fourteen years of age, brown of hair and eyes, and ruddy of cheek, who bore so strong a resemblance to the captain that you hardly needed to hear the latter call him "Hal, my boy," and to note the look of proud affection in his well-bronzed countenance, to know whose son he was. The two were certainly very much alike, both in appearance and in character.

"He's a regular chip of the old block," the captain would say when telling some incident that illustrated Harold's fearless spirit or tenacity of purpose—two qualities that were a better inheritance for him than stocks or mortgages.

The affection felt by the two toward each other was of peculiar strength.   Harold was an only child, and his mother had died when he was but a little fellow.

He had been left to the care of a childless aunt, who, though affectionate and anxious enough, knew nothing about boy nature, and failed entirely to understand her vigorous, enterprising charge.   He therefore not only missed a mother's sympathy and patience, but never came to feel at ease in his aunt's house, everything there being too quiet and precise for his vivacious ways.   And so the brightest periods in his life were when his father would return from one of his long voyages, bringing with him wonderful gifts, and, what was still better, causing a temporary suspension of the firm discipline that made his boy's life unhappy.

The advent of the captain wrought a remarkable change in Aunt Etter's demure household.   He was her brother, and they were the only remaining members of their family.   They were, moreover, bound to each other by sharing in a common sorrow, for she had lost her husband not long before Harold became motherless; and now her whole heart was fixed upon this burly, brown-bearded man, who in his turn exhibited toward her an affectionate tenderness that was the joy of her life.

Nothing that the captain would do was wrong in Aunt Etter's eyes. No sanctuary was too sacred to be invaded by him. He might smoke in the usually silent, shaded drawing-room and not provoke a murmur. No rule of domestic discipline was regarded as binding upon him. The meals were arranged at just what hour he pleased. The lights were burned as late as he liked; and altogether the frigid, formal home life was completely broken up, and Harold felt like a prisoner happily released from confinement.

It must not be wondered at, then, if Captain Marling was more to Harold than an ordinary father; for he was a father, mother, benefactor, and liberator all in one, and his son worshipped him as if he were nothing less than a demi-god.

Harold, on the other hand, received from his father not only his own share of love, but standing as it were in his mother's place, had concentrated upon himself the full power of the captain's big heart; for he could hardly move or speak without suggesting the pretty, graceful woman who, ten years before, had passed gently away, leaving him to Aunt Etter's care.

And so these two were all the world to each other, as the saying is ; and now that Harold was making good progress through his teens, the captain found pleasure in laying before him his plans for the future, as well as in relating the history of his last voyage. It was in this way that Harold came to know of the captain's idea, and forthwith to entertain an idea of his own.

Captain Marling's idea was a rather curious one, and marine circles at Halifax were much concerned about it.   Such an enterprise had never been attempted by any Nova Scotian before, and it was therefore, as a matter of course, pronounced quite preposterous and quixotic.   Yet the captain did not look like a man who would hastily enter upon a wild-goose chase, neither did his past record afford much ground for the gossips and others to work upon.   His reputation was that of a prudent and foreseeing though enterprising and daring man, and the almost uniform success that had attended his previous ventures, making him now, at forty-five years of age, a comparatively wealthy man, showed clearly enough that he usually knew very well what he was about.

Nevertheless, in the face of all this, his friends laughed incredulously or argued earnestly, and his enemies sneered contemptuously, when Captain Marling's idea was under discussion. But the captain minded neither the one nor the other. He was in the prime of life. Every faculty of mind and body was at its best. He had made ample provision for his boy in case of disaster falling upon himself, and now he proposed to indulge in the fulfilment of a desire that had been with him ever since he first took to the sea.

Unlike many other sailors, he had always been fond of reading. His chest had never been without its well-thumbed volume, and somehow or other these books had generally related to the Arctic Regions. The moving stories of Sir John Franklin, of Dr. Kane, and Dr. Hayes, and of the earlier English explorers who had pitted their lives against the terrors of these " thrilling regions of thick-ribbed ice," in brave endeavour to pierce their mysteries, stirred his soul like trumpet-blasts, and in his youth he had registered a vow to make at least one trip toward the Pole as soon as his circumstances would permit.

That time had come at last. Voyaging north, south, east, and west, circumnavigating the globe over and over again, at first in the employ of others, and latterly as his own employer and master, making money often where others entirely failed, his wealth had steadily increased, until he was able not only to put by as much as Harold ought to need, but to have left over more than enough to gratify his long-cherished ambition to try a tussle with the icebergs of the frozen North.

At the same time he did not by any means propose to make his expedition purely a pleasure trip. He would combine business with pleasure, and to this end had determined upon taking a whaling-vessel for his ship, and seeking to secure as many monsters of the deep as would, by their blubber and bone, pay the expenses of the voyage at least. A shrewd, far-seeing man was Captain Marling, and much as his friends and acquaintances might laugh at his idea, and strive to dissuade him from putting it into execution, he only laughed back at them, saying good-humouredly, "Have your say, my friends, have your say. It relieves your mind and does not alter mine, so nobody's the worse. I'm going up north, whether you be true or false prophets."

Aunt Etter was at first in a quite pathetic state of mental bewilderment. She had such unqualified faith in her brother that she could hardly conceive of his doing anything foolish; and yet between her own

vague, exaggerated notions about the Arctic Regions—
which she imagined had a huge polar bear on every
pinnacle of ice, and mighty whales, wicked sword-
fish, and fierce walrus, as thick as sheep in a pasture—
and the ill-advised gossip of her friends, who poured
into her ever-open ears all sorts of terrifying tales,
she was wrought up to such a state of nervous excitement, that it required all the captain's address and
firmness to keep her in check; and only after much
argument and persuasion was she finally induced to
set her mind at rest and cease worrying.

The captain's idea in brief was as follows:—He
had purchased in Dundee, Scotland, a fine steam-
whaler, with all her equipments. He had also engaged a full complement of harpooners, boat-steerers,
and line-managers, taking the utmost care to select
men of good repute; and these, with the engineers,
carpenters, and ordinary seamen, made up as fine a
crew as he had ever had under him. His first mate
and several of his crew were men who had sailed
with him for years, being Nova Scotians like himself;
the remainder hailed from Dundee, coming over to
Halifax in the ship.

The arrival of the *Narwhal* at Halifax created no
small ripple of excitement, and all day long the

wharf to which she was moored received its stream
of visitors, bent on examining as closely as they
might be permitted the first steam-whaler that had
ever been in the port. And the *Narwhal* was well
worth a visit. Solidly built of the very best mate-
rials, her bow sheathed with thick iron plates, and
everything about her speaking of sturdy strength to
resist the deadly embrace of the ice-pack or the cruel
blow of the berg, she presented a very different appearance externally from the ordinary ship or steamer.
Then in her cabins and her hold there was still more
to interest the visitor. The long ranges of iron tanks,
now filled to the brim with coal, but, if the fates
were propitious, to overflow in due time with
unctuous blubber; the strong, swift whale-boats carefully stowed amidships; the arsenal of guns, harpoons, lances, and blubber spades, standing in their
racks, polished to perfection, and ready for service
at once should such a miracle happen as that a whale
should make its appearance in Halifax Harbour; the
comfortable forecastle and cozy cabins, the powerful
engines, well equal to the task of driving the *Narwhal* through opposing fields of ice—all these proved
subjects of lively interest, and to no one more than
the captain's son.

(409)

From the day the *Narwhal* steamed up the harbour and glided into her berth at the wharf, Harold had forsworn all other amusements in her favour. Unless his father wished his company elsewhere, he spent all his free time on board the steamer, until presently there was not a nook or cranny of her hold or cabins that he had not explored, not a mast that he had not climbed, not a harpoon whose quality he had not tested by the familiar process of breathing upon its polished surface. Being the captain's son, the crew, from the first mate down, were naturally very good to him, and many a thrilling story he heard from Red Angus, the big harpooner, or from Colin, the boat-steerer, of tough encounters with mighty whales, or sharp tussles with polar bears. And the more he saw and heard, the more his idea took possession of him, until he became practically a boy of one idea. It was with him awake or asleep. It filled his dreams by night, and drove his lessons out of his head by day. Had he stopped to think about it, he would certainly have said that his happiness depended entirely upon his being permitted to realize his dreams.

But ah! just there was the rub. The prospects of his having his wish were not at all bright, for both his father and aunt seemed strongly opposed to it. The grounds of Aunt Etter's opposition were simple enough.

"No, no, John.   I can't abide the thought of poor little Harold going up amongst the whales and bears and Esquimaux.   It's all well enough for big men, but it's not the place for boys.   Why, John, I lie awake at night picturing to myself our darling boy frozen stark and stiff on an iceberg, or maybe torn to pieces by a polar bear, and it just puts me all of a quiver."

"Oh, well, you needn't take so gloomy a view of it as all that," answered the captain reassuringly. "If he came with us, I'd take good care that he was exposed to no needless risks.   It's not *that* I'm thinking of.   But it seems to me a pity to take him away from school when he is doing so well there ; and there's no telling how long we may be away.   Perhaps only one year, perhaps two or even more ; and yet I must say it goes against my heart to refuse poor Hal.   He seems so in earnest about wanting to come with us."

"I trust, John, you have sufficient strength of mind not to allow Harold's coaxings to change your opinion," said Aunt Etter, rather stiffly, for she began to suspect her brother of showing some signs of wavering.

Indeed, it was very hard for him to be firm and decided. From the first mention of his plan to Harold, the boy had pleaded to be permitted to accompany him; and after the *Narwhal* arrived, and her many interesting features had been all examined, Harold's earnestness rose to fever pitch, until at length one day, when the captain was feeling rather irritable any way, his son's importunities caused him to turn upon him sharply, saying, " Now, Hal, that's enough about it. Not another word, or I shall send you off to the country until the *Narwhal* has sailed."

Hal looked up at his father with an expression of surprise and pain that went right to the captain's heart. His eyes filled, his lip trembled. He seemed upon the point of bursting into a flood of tears. But by a noble effort he controlled himself, and with drooping head turned dejectedly away. Henceforth he said not another word about going in the steamer, but his sad face, his quiet ways, his failing appetite, showed plainly enough how deeply he felt. The captain did his best to cheer him up by being boisterously cheerful, and seeking out all sorts of diversion for him; while Aunt Etter nobly seconded his efforts by preparing toothsome dishes to tempt his indifferent appetite. But all to no purpose, for Harold refused to be comforted by such means.

In the meantime the final preparations for the fitting out of the *Narwhal* were going rapidly forward. Great stores of beef, pork, biscuits, flour, potted meats, canned vegetables, and other food, sufficient for two whole years on full rations, were carefully stowed away; the coal-bunkers were gorged with black diamonds, and the blubber-tanks filled with the same grimy material; the water-tanks one after another received their charge of precious fluid, and there was little left to be done.

Still Harold opened not his mouth. Day after day he went down to the steamer immediately he was free from school, and the kind-hearted sailors, who knew of his desire to go with them, and of his father's refusal, tried to cheer him by telling him to "Never mind!" the captain would take him next time for certain, and promising to bring him all sorts of trophies from the North—a young seal to pet, a walrus's tusk, a bear's paw, and so forth. Harold thoroughly appreciated their kindness—although indeed it only added to his heartache—and felt bound to seem somewhat brightened by it; but all the same his eager longing lost none of its force.

At length the day of departure drew near. On the morrow, if the weather were favourable, the *Narwhal* would set forth. Poor Harold moved about, the very picture of a disconsolate boy. He had two causes for being miserable. The parting with his father was always a great trial to him, and it seemed now particularly hard to bear, when in all probability he would be longer away than ever before; and then there was his own deep disappointment in addition. The last evening the little family of three sat down to tea together, Harold seemed unable either to say a word or eat a bite, although the captain did his best to be merry, and Aunt Etter spread the table with her most delicious dainties. Many a troubled glance did his father give him across the table, and more than once an expression of half-formed resolve flitted across his face that might have given Harold some comfort could he have caught its meaning.

Much earlier than his usual hour, Harold bade the captain an affectionate good-night, and went off quietly to bed, his father saying as the door closed upon him, " Poor little boy ! I hate to refuse him."

Captain Marling remained talking with his sister until a late hour, and then went off to his room. As he passed Harold's door he could not resist the impulse to go in and take a look at his darling boy. Harold was asleep, but there were signs of recent tears upon his pale face, and he stirred uneasily, like one whose slumber was far from sound. Never before had he seemed to the captain to look so much like his dead mother. The big, strong man fairly started at the striking resemblance.

Presently Harold's lips moved. His face took on an expression of passionate entreaty, and he murmured faintly, yet clearly enough to be heard,—

"O father, please take me. I'll be so obedient. Please, father, please do."

Then the light shining in his face awakened him, and his big brown eyes opened wide. In an instant he sprang up in his bed, threw his arms around his father's neck, buried his face in the brown beard, and sobbed out piteously, "Take me with you, father; take me with you, or it will break my heart!"

CHAPTER II.

OFF FOR THE NORTH.

HAROLD'S passionate appeal proved irresistible. Like a dike that the sea has undermined, the captain's resolution had been weakening daily. Not only for Harold's sake, but for his own, he longed to have the boy with him on the adventurous voyage he was about to undertake, and only the desire to keep him a little longer at his school, and the fear lest perhaps he was still too young to pass unscathed through the contaminating influences of shipboard life, had enabled him to stand firm hitherto. And now these defences were swept away with a rush there was no withstanding. Clasping Harold tightly to his broad breast, he said tenderly, " You shall go, my dear boy. You shall go with me. I can't go without you." And his eyes glistened unmistakably as he spoke.

Harold lifted his head, looked straight into his

father's eyes, while a look of inexpressible joy irradiated his face; and then, with a glad cry of "You dear, dear father!" kissed the captain with all his might, and the covenant was sealed.

Aunt Etter was of course greatly shocked when her brother announced his sudden change of mind in reference to Harold. "Really, John," she said, quite sternly, "I gave you credit for more strength of mind. You know very well that it would be much better for Harold to remain with me here than to be thrown into association with the rough men on your steamer. And then just think on all the dangers to which he will be exposed—and he so young, and his education not complete!"

"Pitch into me as hard as you please, Martha; I daresay I deserve it. But I can't help it, all the same," replied the captain good-naturedly. "The boy is dying to go, and I'm dying to have him, and that's an end of it."

"Well, John, of course you're not bound to take my advice in the matter," retorted Aunt Etter; "but if anything happens to Harold, my conscience shall be clear, at all events." And having thus eased her mind, like the sensible woman she was, she said no more, but proceeded to get Harold's outfit in readiness with as much zeal and loving thoughtfulness as if she heartily approved of his going.

The sailing of the *Narwhal* was delayed for a day, in order that Harold might be properly supplied with everything he could possibly require in the way of clothing and comforts; and then at last, on the morning of the 2nd of June, with the sun shining down from an unclouded sky, a gentle breeze rippling the blue bosom of the harbour, and Sam, the black cook, playing "The girl I left behind me," with the skill of a "negro minstrel," on a concertina, the *Narwhal* glided gracefully down the harbour, while Aunt Etter waved a last farewell from her cottage behind the Battery, with the white cloth snatched up from the breakfast-table.

As the throbbing engines drove the whaler onward with increasing speed, and the houses of Halifax lost their individual distinctness and blended into one gray mass that climbed the hill from the harbour to the citadel, Harold felt as if there could be no bounds to his delight. Never before in his life had he felt so happy. His father was too busy to give him any attention; but that didn't matter, for he was just as busy himself. Now he would rush to the bow to watch the waves curling back in foam from the steamer's iron prow ; then he would dive down into the hot engine-room to look at the movements of the mighty machinery ; and next he would, with telescope at his eye, be trying to make out Aunt Etter's trim cottage, and to see if she were still waving the white table-cloth.

Straight out into the ocean went the *Narwhal*, until, having put a sufficient breadth of blue-green sea between herself and the rocky shore, she turned almost due north, and proceeded onward with the Nova Scotian coast at her left, and the great, glorious ocean stretching away indefinitely on her right. There was a fine fresh westerly breeze blowing, which increased in vigour as the morning moved toward mid-day, and presently the *Narwhal* began to pitch and toss in a lively manner, to which Harold was not at all accustomed. He thought it very fine fun at first as the powerful steamer dipped and rose again with the lightness of a duck ; and standing by the taffrail, with the breeze doing its best to blow his cap off his head, he laughed for very joy.

Ere long, however, he began to have a feeling inside as if he had eaten something at breakfast which had raised a row with his digestive apparatus. A most wretched sensation of squeamishness took possession of him, and he was glad to sit down upon the bench beside the cabin skylight, especially as his head was swimming in a very bewildering fashion. He did not feel exactly alarmed at his novel experience, but he did feel very much provoked. All his spring and vigour seemed to have left him, and he became as limp and languid as a jelly-fish. What could be the matter? Was he really going to be ill, and that just at the beginning of the splendid time he hoped to have? The thought was dreadful, and he did so wish his father would come along. But although he knew he could not be far away, he positively had not the strength to get up from his seat and call to him. After sitting there for he knew not how long,—it seemed like hours, it may have been only a few minutes,—his heart gave a bound of joy at hearing the captain's strong voice shouting out, "Hal, my boy, where are you?" With a great effort Harold rose from his seat, and revealed himself.

"Ah! there you are," cried the captain. "I've just been looking for you. First minute I've had to spare since we started. But—hello! what's the matter with you?" he exclaimed, as he noticed Harold's pallid face and limp, forlorn appearance.

Then suddenly, much to the surprise of the boy,

who expected his father to look very concerned and sympathizing, the bluff mariner burst into a laugh and threw himself down beside him, saying, as he put his brawny arm affectionately about his shoulder, "Ha! ha! my lad, and so you're feeling rather miserable in your hold. Well, never mind. We've all been through it. Won't do you a bit of harm. Be all right by to-morrow. Come; I'll take you down to your berth. You'll be more comfortable there." And picking up the big boy as though he were a baby, the captain threw him gently over his broad shoulder, and carried him down to his snug cabin, where he stowed him away in his own berth.

That was Harold's first experience of sea-sickness, and fortunately it did not last very long. He was a pretty wretched individual all that night, and not good for much the following day; but by the evening of the second day he had picked up a bit, and the morning found him as chipper as a sparrow, with a fine appetite for breakfast.

In the meantime the *Narwhal* had ploughed her way past Nova Scotia and Cape Breton, and across the great Gulf of St. Lawrence, and was now nearing the cruel coast of Newfoundland. She had not been making particularly rapid progress, because the wind had been dead ahead, and Captain Marling did not think it worth while to expend a large quantity of coal for the sake of another knot or two an hour.

So soon as Harold had his sea-legs on, he proceeded to make a thorough examination of his surroundings. As regards cabin accommodations, he found himself exceedingly well provided for. The whole stern of the *Narwhal* was taken up with a large saloon, surrounded by a number of state-rooms, the saloon being lighted by a skylight, which made it very bright and airy. The largest and best state-room was, of course, the captain's, and it contained two berths, the upper being assigned to Harold, who therefore had the comfort of knowing that his father was sleeping beneath him and within easy reach.

The captain's cabin was somewhat larger than an ordinary state-room, and fitted up with every possible convenience no pains or money having been spared to make the *Narwhal* a comfortable home. Not even a bath had been forgotten, for what seemed like a luxurious arm-chair proved, on being stripped of its trappings, to be a huge bath-tub that would have contented any Jonathan or John Bull. There were chests of drawers, of rich dark mahogany with bright brass handles, set cunningly into the corners; there were racks of books fixed firmly upon the walls, and containing scores of volumes dealing principally with life and adventure upon the high seas and up in the frozen North ; there were pictures, telescopes, guns, revolvers, and hunting-knives artistically arranged ; and altogether it would have been a queer boy who, after a good look at Captain Marling's snug cabin, did not at once pronounce it perfection.

Harold had his own chest of drawers, his own book-rack, on which his Bible, " Robinson Crusoe," and other literary treasures stood in neat rows, and what pleased him even more, his own rifle and revolver, presented to him by his father on the day they left Halifax, with the promise that he would teach him to use them as soon as he had a little leisure.   He was an in-tensely happy boy.   The relief from the routine of school, and from the prim monotony of Aunt Etter's household, combined with the prospect of a long and adventurous voyage with his father, filled him to overflowing with delight.   Even while his sea-sickness was at its worst he felt no inclination to turn back, although his father pretended to think he did, saying to him very seriously as the *Narwhal* was bowling merrily along, " Pretty sick, my boy, aren't you ? Would you like to go back ?   Just say the word, you know, and I'll run into Sydney and start you on your way home."

Harold was too ill to enter into the spirit of a joke, but not too ill to raise himself up in his berth, and to reply, in a tone of indignant determination that was comically at variance with his pallid face, " No, indeed, sir. You'll do nothing of the kind. I'll soon be all right." And then, as exhausted by the effort he dropped back upon his pillow, the captain patted his forehead soothingly, saying, " Of course you will, Hal, my boy. You'll be all right by to-morrow probably. I was only teasing you a bit."

The other state-rooms that opened off from the saloon were occupied by the first and second mates, the first and second engineers, and the surgeon, and these gentlemen composed the captain's mess. Harold soon made up his mind about each of them. The first mate was, like Captain Marling, a Nova Scotian, a bluff old sea-dog, who was never content away from salt water, free from all family ties, and not showing much affection for anybody save his captain, whom he served with a gruff fidelity that knew no wavering. His name was Peter Strum. The second mate presented a very striking contrast to his imme- diate superior; for he was young, handsome, dashing, and popular. To see Frank Lewis arrayed in his best shore-going clothes, sauntering along the main street of Halifax, you might easily have taken him for some dandy landsman who had somehow or other become very much tanned; but the same Frank Lewis at sea, in a howling gale, with a "sou'-wester" tied tight upon his head and oil-skin coat buttoned close, was a very different being; and despite all his dashing dandyism, at which old Strum was very fond of poking fun in his clumsy fashion, Frank had few superiors as a seaman.

The engineers were Scotchmen who had come out in the *Narwhal*. They were quiet, reserved men, whose hearts were on their work, and were very much attached to each other, seeming not to need any other society, although disposed to be sociable enough when Captain Marling or the others made advances to them. Their names were Donald Stewart and Archibald M'Millan.

Finally there was the surgeon, a high-spirited young fellow, fresh from college, and full of pleasure at the prospect before him. He had a brilliant record as a student, and his friends wondered very much that Ernest Linton did not at once settle down to practice. But he was in no hurry. He was both young and wealthy, and proposed to have a spice of adventure before settling down. Accordingly, no sooner had he heard of Captain Marling's project, than he at once offered himself as surgeon on a nominal salary, and was cordially accepted. He was a good-looking, athletic chap, with a fine record for prowess on the sod and cinder-path, and Harold fell in love with him most promptly. Frank Lewis would be his hero, but Dr. Linton would be his chum, and he felt very rich in his friends.

While these necessary particulars have been getting themselves set down, the *Narwhal* has been steaming onward past the Newfoundland coast, and by to-morrow morning will be in the harbour of St. John's. It was a glorious evening, quiet and clear; the sea was aflame with phosphorescence, and Harold, fascinated by the sight of waves of molten gold rolling back from the steamer's prow, had been a long time at the bow, leaning over the gunwale. The deck was almost deserted, all being below in saloon and forecastle, except the look-out on the other side of the ship from Harold, and the second mate, pacing the poop with steady step and watchful eyes, whistling softly. The forehatch happened to be left open. It lay in the deep shadow cast by the funnel, and presently something strangely like a human head was lifted above the combing, a pair of keen, anxious eyes took a hasty glance about, and then the head disappeared. After a moment's interval it was lifted again. This time it stayed longer, and the eyes caught sight of Harold's form outlined against the bulwarks. The look-out changed his position a step or two, and instantly the head vanished. Once more it was raised, and this time the coast seemed clear. The look-out was motionless, the mate was at the far stern, and only Harold was in sight. As silently as a snake a queer dark form crawled up from the hatchway and crept noiselessly toward the unconscious boy at the bulwarks. Skilfully keeping in the darkest shadows, it got right behind him, and then paused, for between him and it there was a broad band of moonlight it dared not cross. One, two, three, four, five minutes passed, and there was no further movement. Then Harold, saying aloud, "Dear me! time to go to bed," turned away to go toward the cabin. In so doing he passed by the shadow wherein the mysterious form cowered tremblingly. Then like a flash an arm was thrust out, and a thin hand was laid upon his arm, while a weak, pitiful voice said, in tones of agonized entreaty, "For the love of God, sir!"

CHAPTER III.

A PECULIAR PROTÉGÉ.

WITH a great start of surprise Harold turned upon the speaker. He beheld before him, somewhat obscured by the heavy shadows, a curious-looking being. It was a boy, seemingly of about his own age, but a difference of circumstance had wrought a wonderful difference of appearance. While Harold was ruddy of face and sturdy of figure, this poor creature was wan and pinched, with garments hanging about a frame that was not very much better than a skeleton. Yet his face was not unpleasing. Sharp and starved as it looked, its expression was in no wise vicious, and the passionate intensity of entreaty it now expressed would have touched a far harder heart than Harold's. Noticing the latter's hesitation, and fearing lest he should call out, this unexpected apparition made as though he would draw him back into the shadow, saying, in the same piteous tones, "For the love of God, sir, listen to me."

"What is it? what do you want?" asked Harold, not unkindly, unconsciously yielding to the other's influence, and stepping aside into the dark nook from which he had so suddenly emerged.

"Please, sir, I'm starving, and I'll die if I don't soon get a bite to eat. I've had nothing to eat but a bit of a loaf since we started," answered the boy, whose whole appearance gave such strong support to his story.

"But who are you, and what are you doing here?" queried Harold, looking hard at him.

"Please, sir, my name is Patsy Kehoe, and I'm a stowaway: that's the truth of it, and it's no use a-denyin' it." And Patsy hung his head as if he dared not look his questioner in the face.

A stowaway! The name touched Harold's sympathies at once, and made his heart beat more quickly. How often he had read about these daring chaps who would hide away in dark, noisome holds, and there endure all sorts of suffering until the time came to reveal themselves! But until now he had never seen one, and he regarded Patsy with a keen curiosity not unmingled with admiration. Quickly noting the friendliness of his look, Patsy spoke again: "You won't give me away, sir, will you?"

"Give you away? How do you mean?" asked Hal, not quite understanding him.

"You won't tell the captain on me? He'll put me off at St. John's if you do," explained Patsy.

"Why, the captain's my father," returned Hal, with a proud ring in his voice, "and he'll let you stay if I ask him."

"Oh, please, sir, don't ask him, sir! He'll put me off sure, if he finds me out. That's the way they did with Teddy Shea only last summer, and he had to find his way back as best he could."

"Father's too kind to do that," said Hal confidently. "But, anyhow, I won't tell him if you don't want me to. What do you want of me now?"

"Won't you, please, sir, get me a bit to eat? and then I'll crawl back into the hold again."

"But surely you won't go down into that dark, dreadful place again!" exclaimed Hal, with a shudder.

"I will, please, sir, until the ship's got past St. John's, and then I'll just take my chances with the captain."

"Well, stay here till I get you something to eat, any way," said Hal, going off toward the cabin.

He had the freedom of the ship, and no one ever thought of questioning his movements; so that it was easy for him to go into the steward's pantry—which he was glad to find deserted—and take anything he pleased from it. Hastily filling two pockets with biscuits, and snatching up a bone of ham that had a good deal of meat still left upon it, he hastened back to the bow.

Patsy's eyes flashed at the sight of the ham and biscuits, and with a fervent "Lord love you, sir!" he proceeded to stow them away as best he could in his tattered garments. It made Hal's heart ache to see the poor little fellow's trembling eagerness, and when he ventured one more request, he was only too glad to fulfil it.

"Could you get me a sup of water, sir? I'm powerful dry," begged Patsy.

"Of course I can," responded Hal promptly. He instantly thought of a tin can with a cover to it that he had seen that evening in Sam's galley, and went off after it. Sam was not present to protest, and securing the tin, Hal filled it full with water and returned to his newly-found *protégé*.

"God bless your kind heart! It's nothing less than an angel that you are!" ejaculated Patsy.

"And now, sir, I must be after getting back to my place again."

In much perplexity Harold watched the dauntless little fellow disappear down the half-open hatchway as softly and silently as a shadow. He felt so sure of his father's permitting the stowaway to remain on board that he longed to run and tell him about Patsy. But Patsy had begged so hard for him not to do this until after the *Narwhal* had left St. John's that to do so now would seem like breaking faith with him. Yet the thought of his passing another night, and perhaps even longer, in some dark corner among the casks and tanks, when he might be snugly ensconced in the forecastle, was very repulsive.

Harold longed to take counsel with some one. But in whom should he confide? He felt as though he did not know any of the men on board well enough yet to trust them with his secret. No; if he could not tell it to his father, he could not tell it to anybody—that was clear. And so, feeling that he had no small burden on his mind, he returned to the saloon, where his father hailed him with a cheery "Well, Hal, my boy; had enough star-gazing? about time to turn in, isn't it?"

"I'm just going, sir," answered Harold. And bidding him good-night, he was soon snug in his berth.

But he was not soon asleep, for Patsy Kehoe kept him awake; and when, an hour later, the captain himself turned in, he was surprised to find his boy's eyes still open. "Hello, Hal! not asleep yet? What are you thinking about?" he asked, in an affectionate tone.

It was only by a great effort that Hal restrained himself from there and then making full confession. But he did manage to keep his counsel, replying brightly, "Just been waiting for you, sir. I'll soon go to sleep now."

Captain Marling threw himself into his berth, and presently his heavy breathing announced that he was off to slumber-land. But Harold still stayed on this side the magic boundary. Two different thoughts were troubling his mind. He could not help thinking of poor little Patsy cowering comfortlessly somewhere in the forehold, while he nestled snugly in a luxurious berth; and then again he felt as though he were doing wrong in keeping Patsy's presence a secret from his father, now sleeping so soundly just beneath him. At length, however, the long-delayed sleep came, and with it dream after dream, in every one of which Patsy figured. Now he was being ordered off the ship by Captain Marling, who looked dreadfully angry, and now he was in command himself, and was ordering Captain Marling to do something; and so it went on, until morning came and found Harold not as much refreshed as he ought to have been by his night's rest.

The *Narwhal* had reached St. John's, and going on deck Harold saw round about him the towering cliffs of this famous harbour, and straight in front the amphitheatre of the city. Here they spent the day purchasing furs, adding some further supplies to their larder, and securing the addition to their crew of two experienced sailors—men thoroughly acquainted with the wild region for which they were destined. In the bustle and activity that filled the day, Harold quite forgot his *protégé*, and it was not until as the sun sank in the west the steamer turned away from St. John's, and passing out through the harbour's narrow gateway began once more to toss about upon the Atlantic's restless bosom, that he bethought himself of the boy in the forehold. When he did think of him, he was impatient for the time to come when he need be burdened with his secret no longer.

The evening was fine and clear, the breeze fresh and favourable, and Captain Marling sent the steamer along almost at her best speed, while he stood on the bridge talking earnestly with one of the men who had joined them at St. John's. He was evidently in excellent humour, for his hearty laugh could be heard now and then ringing out as the sealer would say something that tickled his fancy. Harold noted all this, but still it would not do to interrupt him just then.

The sun vanished, the stars came one by one into their places, the sailors put the ship in order for the night, and then everybody went below except those whose duty it was to remain on deck. Captain Marling took the sealer down into the saloon to consult some charts with him, and all was quiet on board the *Narwhal*, save the ceaseless throbbing of the engines, and the plashing of the waves that seemed to be vainly striving to oppose her onward progress.

Now was the time for action. Going forward to the hatch, fortunately not yet battened down, Harold gently removed a section of the covering, and leaning over into the dark well, called out softly: "Patsy! Patsy! Come!" There was a stir somewhere over in a far corner, a whispered "I'm just a-comin', sir," and presently the thin, sharp face was underneath Harold, and peering up eagerly at him.

"Come along, Patsy; coast's clear," said Harold, in his most encouraging tone.

Without a word, Patsy wriggled up on deck, and

stood beside his friend. He was about as tall as Harold, but wofully thin, and evidently very weak—as genuine an object of compassion as one could well imagine. Harold put his hand upon Patsy's shoulder with a gesture of protection. "Poor fellow! you've had a mighty hard time of it. But you'll be all right before you've been a month with us."

Patsy looked up at him gratefully. "It's sure I am that you'll be kind to me, sir," said he.

"Oh, I won't be the only one. Come along, now, and let father see you," said Harold, moving off toward the cabin. Patsy hesitated a moment. He shrank from going into the presence of the big, bluff, strong-voiced captain, whom he had been watching for days at the wharf before he dared to stow away on board his steamer. . But it had to be done. No matter what the captain might do to him, it could hardly be so bad as hiding in that horrid hold with his life in constant peril from hunger and thirst, not to mention the huge rats that ran over him in the darkness, and stole what little food he had, fighting fiercely together over it afterward. Then again he had the captain's own son to plead for him, and surely that would count for a great deal.

Strengthening himself by these reflections, Patsy

followed Harold along the deck, and up the ladder to the poop. There they encountered the second mate, who was in charge of the steamer.

"What in thunder have you got there?" exclaimed Lewis, as they approached him.

Harold held up his hands warningly. "Hush!" he said. "Don't say anything. It's a poor little stowaway, and I'm going to take him down to father."

Lewis gave a long whistle. "Humph!" said he, "I'm afraid he won't get a very cordial reception there. The captain's down on stowaways."

Poor Patsy gave an apprehensive shudder, but, quite undaunted, Harold answered cheerfully, "Oh, that's all right. He won't be down on this one."

"Perhaps not. You'd better go down and see, any way. He's in a very good humour now too," returned the mate.

Although his heart was beating like a trip-hammer, Harold did his best to look very composed and at his ease as he entered the bright, well-furnished saloon, which seemed like Paradise itself to the city waif following tremblingly in his footsteps. So close did Patsy keep to his protector that Captain Marling, who was at the other end of the saloon, did not notice him when he looked up, and said, "O Hal, I'm glad you've turned up. Come here, and I'll show you on this chart where the *Narwhal* is to go."

Harold turned first red, and then pale, and his voice had a tell-tale quiver in it as he stepped forward, and pointing to Patsy, now fully revealed in the bright light, said, in a tone of assumed gaiety, " I've brought you another passenger, father."

Captain Marling sprang from his chair, a fiery flush suffused his face, and his eyes flashed angrily as he cried, " What's the meaning of this, boy ? What trick have you been playing upon me ? "

The stowaway started as though he would have made for the door; but Harold caught him by the arm, and facing his father fearlessly, replied, in respectful tones, " I've been playing you no trick, sir. If you ask Patsy he will tell you all about it."

" Well, you know I don't want any good-for-nothing wharf-rats on my steamer," returned Captain Marling, in a somewhat milder tone.—" Come here, you," looking crossly at Patsy, "and let me hear what you have to say for yourself."

" Don't be afraid ; just tell him the truth," whispered Harold, encouragingly, in Patsy's ear.

It was a curious scene—the poor, ragged, dirty,

starved, wretched-looking stowaway standing before
the big, brown-bearded captain, under the strong light
of the swinging lamp, and telling his story in his
own simple fashion, the Irish brogue proclaiming
itself in every sentence, while a group of men
gathered about with interested, sympathetic faces.

Patsy's story was a very pathetic one. His mother
was dead, having lived only long enough for him to
remember how good she was to him. His father had
married again, and the step-mother had nothing but
cross words and cruel blows for the child that was
not her own. For years his life had been one of
constant misery, until at last, driven to desperation,
he determined to stow away on board some vessel
that was going on a long voyage; and hearing about
the wharves of the *Narwhal's* destination, he thought
he could not do better than hide himself in her hold
and take his chances.

All this came out by dint of much questioning, and
it was apparent that the recital had due effect upon
Captain Marling's heart. His tone, which at the first
was one of impatience and anger, gradually softened
into one of sympathy and pity, until at last he found
himself saying, as he looked from Patsy's shrunken
form and pallid, pinched face to his own boy's sturdy

frame and ruddy cheeks, " Poor little chap! you have been used hardly, and no mistake."

Then, as if recollecting himself, he drew his face up into sterner lines, and seemed to be considering what action he should take. Harold, impatient for his father's decision, and fearing lest it should be adverse, here interposed, saying in tones of earnest entreaty, while the tears stood in his eyes, " You'll let Patsy stay, won't you, dear father ? "

Captain Marling looked from one to the other of the two boys before him with a quizzical smile. " Humph ! " said he at last. " Not content with making me take you whether I would or no, you now want to force another youngster upon me. Well, I'll tell you what I'll do Patsy can stay; but mind you, if he doesn't prove worth his salt at least, I'll ship him back to St. John's by the first sealer I meet. We've no room for useless boys on board the *Narwhal*. Take him away now, and let him get a good wash and a bite to eat."

There was a murmur of approval at the captain's decision from the listening circle, and full of joy at the result, Harold marched Patsy off to carry out his father's instructions.

CHAPTER IV.

A BATTLE WITH THE ICE.

A THOROUGH wash, a hearty supper, and a good night's rest in a spare bunk there happened to be in the forecastle, effected a wonderful change in Patsy Kehoe; and when, the next morning, Harold made him throw his old rags away and put on in their place a suit of clothes that still had plenty of wear in them, and fitted him tolerably well, he presented so improved an appearance that Captain Marling looked quite graciously upon him, and decided that he should be his cabin-boy, his duties being to wait on table at meal-time, keep the saloon shipshape, and make himself generally useful. This light work suited the boy very well, and being extremely eager to please, and very nimble with both hands and feet, while possessing a good share of brain withal, he soon became well worth his salt, and there was not much danger of his being sent back on the first sealer.

His devotion to Harold was really beautiful to witness. He looked upon him as the good genius that had wrought so wonderful a change in his fortunes. His eyes followed him with the same deep, trustful, loving look that makes a dog at times seem almost human, and he was ever ready to anticipate, if possible, any want of his young champion's—to fetch and carry for him all day long. The time was to come when Patsy would make still greater proof of his gratitude, and when not only Harold, but Captain Marling too, would bless the day that they showed kindness to the little Irish stowaway.

Onward, past the stern, terrible Labrador coast went the *Narwhal*, beset by storms that seemed to have neither beginning nor end, encompassed by ice that covered the ocean's bosom with a tossing, tumbling breastplate, or rose into mighty bergs that towered threateningly about the steamer on every side, or swayed in dense chilling fogs that compelled Captain Marling, impatient as he was to reach Hudson Strait, to take in all his canvas and work the engines at half-speed.

Harold was in a constant state of wide-eyed admiration and delight, tempered with delicious thrills of terror. He had lived a pretty quiet, uneventful life at Aunt Etter's, and the only adventures he had known were such as commonly fall to the lot of a boy in a bustling seaport who is allowed a proper degree of liberty. He had had one or two narrow escapes from drowning; he had upset out of a boat in the middle of the harbour; he had been lost in the woods while out on a trout-fishing expedition; but that was about the sum of his acquaintance with danger, and now he seemed to be in the very midst of it. The mighty ice-king had apparently arrayed all his forces against the *Narwhal* in a determined effort to prevent her penetrating into his dominions. The gigantic bergs by tens, with scores of " growlers," as the little bergs are called, came trooping down before a north-east gale; and there were nights when the berth below Harold's was undisturbed, for the captain would not leave the bridge. Then one day it would be fog so dense that you could not see from one bulwark to the other, and the next a blinding snow-storm that made the look-out's duty anything but pleasant. And all the time there was a heavy and confused sea running, and the *Narwhal* laboured like a weary horse, giving vent to frequent groans that had a weird, life-like sound.

This had been going on for a full week when something still worse happened. Late one evening, the ice, hitherto broken into cakes and floes, set solid to the ship fore and aft, rafting and piling up all around, and for the next ten days the stout *Narwhal* was as helpless as a fly upon a carriage-wheel. Hither and thither, up and down, north, south, east, and west, powerless to disobey the bidding of the fickle winds or changing currents, the steamer drifted in the firm grip of the ice, which seemed as though it would never weary of its plaything. When a change of wind came—and the wind was frequently veering round to some fresh quarter—the pack would open a little, and in hope of escape the engines would be set hard to work. But before a mile had been made, the pack would close again, seeming thus to be playing with its prisoner as a cat plays with a mouse.

The effect of all this delay and uncertainty was very visible on board the *Narwhal*. Captain Marling's bluff cheeriness gave place to anxiety and irritation; old Peter Strum grew more crusty and crabbed, if that were possible; and even Frank Lewis's usual high spirits suffered an eclipse to such an extent that he seemed too much vexed to whistle. The sailors, too, began to murmur and grumble, to throw out hints about theirs being an unlucky ship, and to relate what they conceived to be omens in support of their forebodings. But in the midst of it all there was one upon whom these untoward circumstances had not the slightest effect. His face was always bright and sunny, his voice full of cheer, and his step buoyant. All day long while at his work he whistled or sang merrily to himself. His life seemed one unshadowed dream of delight. This was Patsy Kehoe, late dirty, ragged, starving stowaway, now clean, well-clothed, well-fed cabin-boy. "Bless the youngster!" the captain would say, as Patsy's blithe whistle came up like the warble of a bird from the pantry, where he was busy helping the steward— "bless the boy! he's the only one on board that's got any spirit left. We might do worse than take a leaf out of his book."

One evening after dinner, during which meal Patsy had been waiting upon the table with a face as full of brightness as his motions were quick and his fingers deft, Captain Marling called him to him. "Look here, Patsy," said he: "how is it you're so 'chirpy' when everybody else is so glum?"

"I don't know, sir," answered Patsy, with a respectful tug at his forelock. "S'pose it's because I'm having such a good time."

"Oh ho!" laughed the captain. "So this is what

you call having a good time. I wonder what Peter Strum would say to that."

" Well, you see, sir," said Patsy, feeling as if he were in a certain measure called upon to explain himself, " it's just this way. You've let me stay here, sir—God bless your kind heart!—and there's plenty to do, and plenty to eat, and a nice corner to sleep in ; and then Master Harold has given me these nice clothes, and so it's better off I am than ever I was in my born days."

" Good for you, Patsy, my boy," smiled the captain. " You've got sense enough to know when you're well off, at all events. Stick to that, and you'll save yourself many a heartache."

Day after day dragged itself wearily along from sunrise to sunset, and still the *Narwhal* was held in durance vile. Then one morning a wild storm burst upon her from the north-east. The wind raged and screamed through the rigging, the heavy clouds hung so low that it was almost dark at mid-day, and the sea-birds flew frantically about the steamer, as if they looked to her for refuge. At first so vast was the pack, and so firmly was the *Narwhal* imbedded in the very heart of it, that the fiercest gusts hardly produced a tremor in her mighty frame But presently the pack began to break up at the edge most fully exposed to the storm. With surprising rapidity floe after floe was detached, with a report that sounded thrillingly like thunder, each floe in its turn taking up the part of battering-ram against the inert, defenceless pack, until at length the whole field was shattered into fragments, which, now no longer stationary, tossed and tumbled ceaselessly at the bidding of the storm. It was then that the real peril of the *Narwhal* began. Captain Marling wanted Harold to remain below; but he begged so hard to be allowed to stay on the bridge beside his father, that he was permitted to have his own way. From there he had a full view of the struggle now going on between the forces of nature in their destructive fury and the efforts of man to cope with them.

The steamer had been, so to speak, prepared for action. Her hatches had been battened down securely, all loose ropes coiled snugly in their places, and the deck cleared of everything not needed. Every man was at his post. On the bridge were the captain, the first mate, and one of the sealers who had joined at St. John's, often conferring together as to the best thing to do. On the poop were the second mate and the second engineer, the latter having come up from the hot engine-room for a breath of cooler air, while his senior officer took his place below ; while in the waist of the ship and at the bow the men were gathered in little groups, awaiting their captain's command.

As bravely and obediently as though she were a thing of life did the gallant steamer struggle onward. Harold, hugging tightly the rail of the bridge, a heavy tarpaulin hat tied fast upon his head, and a thick oil-skin coat protecting his body, hardly knew whether terror or fascination had more control of him ; and the scene before him was certainly well calculated to inspire boy or man with both emotions to no small degree.

The storm had now reached its height. The furious wind turned the tight-strained rigging into a huge Æolian harp, and played upon it tunes that sounded like the wild wails of tortured spirits. The steamer groaned through every timber as she rose and plunged in the mighty waves. But grandest and most terrible of all were the ice-floes—some large enough to have borne the *Narwhal* upon their backs as lightly as a powerful horse does its rider ; others no bigger than one of the whale-boats hanging in the davits. They were fully five feet thick, of hard Arctic ice—a portion of Jack Frost's last winter's crop; and driven by the impetuous wind, they tossed about on the tumultuous waves, crashing into one another with splintering collision, or battering the steamer's ironclad hull with dull booming shocks that made her tremble from stem to stern.

The keenest vigilance was required to avoid collisions with immense floes that might have done serious damage, or to elude the fatal nip of icy monsters charging madly upon one another. The steward crept cautiously up on deck to announce that dinner was ready, but nobody heeded him. The dinner might wait, but the storm would not, and no man could be spared from his post. Not that there was much to do, save to look on at the tremendous conflict; but who could tell when the best efforts of all might be required?

Finding the wind unbearable on the exposed bridge, Harold made his way back to the poop, and sought out a sheltered nook beside the mizzen-mast. He had not been there long before Patsy, relieved from his duties below, came creeping along the decks, holding fast to whatever was at hand, and Harold, delighted to see him, called out, "Patsy! Patsy! come here!"

Patsy at once steered towards him, and the two boys nestled sociably together in their corner, whence they could look out upon the storm, and listen to the grinding and crashing of the ice-floes.

" It is dreadful, Patsy, isn't it ? " said Harold, in awe-struck tones. " I'm so afraid; aren't you ?"

" Faith, it is a bad storm, Master Harold, and that ice just looks as if it was achin' to smash us up altogether," replied Patsy; " but we'll be taken care of, and I'm not afraid."

" Just think, though, Patsy: if one of those big cakes of ice was to make a hole in the steamer's side, and she was to sink, what would become of us then ? We couldn't take to the boats, and we'd just have to drown," went on Harold, who felt disposed to take rather a gloomy view of the situation.

" Arrah, now ; don't you be thinking of that at all," returned Patsy, putting on a bright smile. " Sure, nothin' of the kind's goin' to happen at all; the steamer's too strong for any piece of ice to make a hole in her."

" I don't know about that," said Harold, shaking his head doubtfully; " I've read of it happening, in the books."

" Was it only in the books it happened ?" retorted Patsy, who seemed determined to be cheerful. " Then, sure, we needn't bother ourselves about it, for it's not in the books we are."

While they were speaking, Frank Lewis came

along the deck, and noticing the boys in the corner, approached them, saying, " Well, boys, don't you wish you were safe at home again ?"

" I don't know but I do, sir," answered Harold.

" Faith, then, and I don't," cried Patsy. " I'd rather be on board this steamer in the biggest storm that ever was than be at my home again."

Lewis smiled at the speaker, understanding well his meaning. " You know when you're well off, don't you, Patsy ?"

" I think I do, sir; and I'm well pleased to stay, storm or no storm."

The afternoon wore away without any apparent abatement in the violence of the gale or the quantity of the ice, and with the approach of night anxiety on board the *Narwhal* increased. The steamer had so far stood the ordeal splendidly. The pumps had been frequently sounded, but no sign of a leak was discovered. The engines had been doing their work as smoothly and steadily as though there were no storm, and beyond some slight breakages in the saloon and pantry, no damage had been reported.

All this had been made possible only by the exercise of the utmost care and skill on the part of the captain and his assistants. So admirable was Captain

Marling's seamanship that even Peter Strum hardly ventured a suggestion. All day long he had stood upon the bridge, watching every movement of his gallant ship, and of the ice that seemed to be conspiring for her destruction. Just before dark he left his post for a few minutes to snatch a hurried bite of food, and then returned there to spend the rest of the night.

At supper-time those who could be spared from the deck gathered at the table—Harold, in spite of his uneasiness, finding himself with a keen appetite. They were just in the midst of their meal, when there was a heavy concussion under the steamer's stern. The propeller seemed to cease its revolutions for an instant, and then start again with a rush. But, clearly enough, there was something wrong. Instead of running with the regularity with which it had been running all day, its jerky, spasmodic action told only too plainly that serious damage had been done in some way.

The chief engineer sprang from the table with a troubled face and hastened off to the engine-room. Frank Lewis hurried on deck, while the surgeon and Harold remained in the saloon, anxiously awaiting an explanation of what had happened. What seemed a long time to them passed without any one appearing;

and Dr. Linton was just about to go up on deck, when Lewis ran down to get something from the cabin.

"Hello, Lewis, what's happened ?" cried Dr. Linton.

"A pretty bad business," replied Lewis, whose countenance showed much concern. "The propeller's broken."

"Propeller broken!" exclaimed the doctor. "How do you mean? Broken right off?"

"Oh no, not so bad as that," answered the mate, unable to restrain a smile. "But one blade must be gone, any way; and we can't tell till morning what other damage has been done."

Harold's heart sank within him at these words, and Dr. Linton looked very grave. There was, indeed, plenty of cause for great anxiety. The situation of the steamer was perilous in the extreme. To have faced such a gale with everything in order would have been task enough during the long hours of darkness; but to struggle with tempest and darkness together when in a semi-disabled condition! Little wonder if Captain Marling's well-bronzed countenance took on a haggard look that did not seem at all natural, and if Peter Strum was overheard continually muttering unintelligible things, that may perhaps have been prayers, as the anxious night wore slowly away.

CHAPTER V.

AT NACHVAK.

THE gray morning dawned with lingering reluctance, and found the *Narwhal* still struggling stoutly and successfully with her besetting dangers. As the light grew stronger, Captain Marling, to his great joy, discovered that the ice-pack no longer covered the sea as far as eye could reach, but that off to the north-west clear water could be discovered. Moreover, the storm had certainly decreased in violence, and the outlook was altogether more hopeful.

"We'll soon be out of mischief, Peter. We ought to make that clear water in the course of an hour," said he, in a cheerful tone, to the old mate.

"Ay, ay, sir, it's all right. But we've had a hard night of it," responded Strum, with the semblance of a smile.

Within the hour the open water was safely reached, and all danger from the sea became a thing of the past. The wind continued to go down during the morning, and preparations were made to repair the propeller. This would be no easy task. Captain Marling was, of course, too wise a seaman to start upon such a voyage without taking the precaution to provide himself with some spare blades for the screw; but it was one thing to have them stowed snugly on board, and another to put them into use.

Harold watched the proceedings with intense interest and a constant shower of questions. So soon as the wind had sufficiently abated and the sea become less turbulent, a hoisting-gear was arranged at the stern. Then, under the direction of the chief engineer, the huge heavy screw was slowly lifted from its place, and, with many a "Heave-ho" and "Easy now," swung up on the deck, where the damage done to it at once became apparent.

One of the blades had been broken short off, and it was a wonder to all that it had served its purpose so well as it had during the dark hours of danger. Two hours of hard work and a new blade restored the screw to its original condition. Less than half that time was sufficient to replace it at the stern; and then, with everybody on board feeling that a great burden had been lifted from them, and that they had glad hearts and free, the *Narwhal* went bounding over the waters, steering directly for Hudson Strait.

So soon as this had been successfully accomplished, Captain Marling, thoroughly tired out, went down to his cabin for the rest he so greatly needed, and did not appear again until supper-time. Then he came out in excellent humour, and had many questions to ask of Harold, who was in close attendance upon him.

"What do you think of a sailor's life now, Hal? Not much fun about it in a storm, is there?"

"No, indeed, sir," replied Harold promptly. "I'm mighty glad the storm is over, and hope there won't be another in a hurry."

"There's no telling, Hal; there's no telling," said the captain, shaking his head. "Plenty of storms up in these regions. Perhaps," he added, with a sly look at his son, "you wish you were back in Halifax again."

"Not a bit of it, sir," protested Harold. "I was a little frightened last night, but I'm all right now, and it'll take more to frighten me next time."

"That's a good way of looking at it, Hal," said the captain, with an approving smile. "Just keep along that line, and you'll soon get as hard to scare as old Peter himself."

Harold blushed with pleasure at his father's commendation, and firmly resolved in his mind that he would keep along that line, and that even though he should never become quite as hard to scare as old Strum, who seemed to be made of sole leather, he was so tough, still he would follow his example closely.

The *Narwhal*, under a full head of steam, and with every stitch of canvas set, made great headway northward during the day. The little ice that there was scattered over the face of the ocean offered no opposition to her progress, and ere the sun sank to rest behind the lofty cliffs that lined the shore, she had cast anchor at Nachvak Bay. There Captain Marling intended to procure, if possible, an Esquimau, who might accompany him for the remainder of the voyage as interpreter.

Harold was highly delighted at the prospect of a run on shore, especially as it would probably afford him his first glimpse of a genuine Esquimau; and long before the steamer came to anchor, he had obtained permission from his father to accompany him when he went on land. This the captain did not do that evening, as it was almost dark before the *Narwhal* was anchored to his satisfaction ; consequently Harold had to restrain his impatience as best he might until morning. They had a visitor, however, whose coming in some measure consoled him—namely, the chief factor of the Hudson Bay Company's post, which gives to Nachvak what little importance it possesses.

This gentleman, who might be said to have been monarch of all he surveyed—for certainly his right there were none to dispute, the gentle Esquimaux being entirely subject to him—proved to be a stout, full-bearded Scotchman of about middle age, with the manner of one accustomed to do and say pretty much what he pleased. He came out in his boat just about dark, and despite his important bearing, Harold thought him of little interest in comparison with the quartet of oarsmen that composed his crew. These were unusually fine specimens of Esquimaux—four short, squat, dark-skinned, black-haired, brown-eyed, flat-nosed individuals, who seemed to be very princes of good-humour. There was not a line of care or hint of temper on their fat faces; and when, having secured the boat, they clambered awkwardly but fearlessly up the rope ladder at the *Narwhal's* side and stood in a little group upon the deck, Harold thought them the very oddest-looking fellows he had ever seen.

He at once went up to them, and was not a little proud to find himself almost a head above the tallest of the four.

"Good-morning. Glad to see you," said he, in his most gracious manner.

The Esquimaux grinned broadly, and after looking at one another said something together that was evidently intended as a reply to the salutation, but just what it was Harold could not for the life of him make out.

"Can you talk English?" he asked.

The dusky visitors grinned again, and, after another look at each other, shook their heads.

"That's a pity," said Harold; "for I certainly can't speak your language."

Suddenly a happy thought struck him. Though they couldn't talk, they undoubtedly could eat; so bidding them "stay there, he would be back in a second," he darted off to the saloon, and presently re-appeared bearing a plateful of biscuits. He could not have made a better choice. The moment the Esquimaux saw the biscuits their eyes gleamed with delight, their grin extended well-nigh from ear to ear, and as soon as the plate was within reach they simultaneously made a grab at it, with the result that several biscuits were knocked off the plate to the deck, where they scrambled for them with all the eagerness of street arabs scrambling for pennies.

Harold laughed heartily at their naive manners, and Patsy appearing at this moment, he called him over to share in the amusement. Their eager scrambling for the biscuits had given him an idea which he forthwith proceeded to put into execution. Withdrawing a little way from the Esquimaux, who were still standing by the bulwarks, he held up a biscuit, and calling out, "Here! Catch! The first man that gets it keeps it," threw it into their midst.

Not for an instant did the natives hesitate. They may never have played that sort of game before, but they knew exactly what to do, all the same, and with the agility of four expert football-players they made a dive for this biscuit. The struggle that ensued was very funny, and the two boys laughed until their sides ached as biscuit after biscuit was thrown with the same result, the Esquimaux thoroughly entering into the spirit of the thing, and manifesting the utmost good-humour. Nor were the boys the only spectators. The entire crew soon turned out to witness the sport, and were joined by the second mate and surgeon; so that when Captain Marling came up on deck with the factor, he found a noisy crowd gathered amidships, which he quietly joined, and, with his visitor, enjoyed the fun as much as the others.

Harold had just despatched Patsy for a fresh supply of biscuits, and the excitement was at its height. Tossing the biscuit into the air, he would shout, " Now, then—jump for it !" and jump for it the four fat " Huskies " would, knocking the biscuit on the deck and rolling over one another in their eagerness to get it. The one who succeeded in securing the prize would then stuff it into the capacious bosom of his shirt, and be ready for a fresh struggle.

By the time the second supply of biscuits was exhausted the factor wished to return to the shore, and in vast good-humour his swarthy oarsmen took their leave of the white visitors who had treated them so handsomely.

The following morning proved gloriously fine, and Captain Marling, immediately after breakfast, ordered his gig to be put in the water, that he might be rowed ashore in state to return the factor's visit. Harold, as a matter of course, accompanied him, donning his best shore-going clothes for the occasion, and in high spirits at the prospect of a run on dry land ; for if it must be told, he had already begun to find the confinement of shipboard life not a little irksome, and a change, however brief, was wonderfully welcome.

Nachvak could not boast of many lions wherewith to entertain its visitors. Half-a-dozen low, strong wooden buildings, gathered into a sort of square, constituted the Hudson Bay Company's post, and besides that, at a little distance off, a cluster of the picturesque-looking skin tents in which the Esquimaux live during the hot months of their brief summer were the only signs of human habitation. Of vegetation, except some scanty patches of moss and lichen, there was none; but here and there, in sunny, sheltered nooks, tiny green leaves might be found thrusting their way through the unpropitious soil.

The factor—who rejoiced, by the way, in the good Scotch name of Donald M'Tavish—greeted his visitors very cordially, and escorted them at once to his quarters, where Harold had the opportunity of going about on his own account, while the elders partook of some refreshment in the factor's best parlour. He was soon an object of lively interest to a group of Esquimau children, who followed him about like a pack of dogs, watching his every movement and chatting to one another. They kept at a very respectful distance, and Harold, wishing to be friendly, made several attempts to overcome this reserve; but his efforts resulting only in a very sudden and complete scattering of his dusky followers, he gave it up as a bad job.

Presently, however, he had better fortune; for after he had made the rounds of the post, looking into the storehouses, now filled with furs awaiting the coming of the annual ship, and taking a peep at the men's quarters, which seemed rather close and stuffy abodes for summer-time, although no doubt just the right thing for the bitter days of mid-winter, he went over to the Esquimau village. He no sooner approached it than a fat little man came toward him, grinning from ear to ear, whom Harold, with some little difficulty, recognized as one of the factor's boatmen. He was very glad indeed to see him, and holding out his hand, gave him a cordial " Good-morning."

The Esquimau looked at the outstretched hand as though he expected to see something in it. Then finding it was empty, he took hold of it in both of his and pressed it to his bosom. He was a jolly-looking little man, quite three inches shorter than Harold, but considerably broader, and the boy could not help thinking to himself that if it came to a hand-to-hand struggle, the " Huskie " might prove a very tough customer. Harold wished very much that he had taken some lessons in the Esquimau language before coming up north, for it did seem so stupid to be standing there *vis-à-vis* with this pleasant-faced native, and to all appearance trying to outdo him in grinning.

One can always fall back upon the sign-language, however; so pointing at the cluster of tents, Harold called out loudly, as if his hearer was somewhat deaf, " I'd like to see your tents. Come along and show them to me."

Whereat the Esquimau enlarged his smile by way of indicating that he understood, and at once waddled off toward the tents, with Harold following close in his wake. There were about twenty tents gathered together in an irregular group, and made apparently of very poor parchment, the actual material being seal-skin, with the hair carefully scraped off. Each tent was the abode of a family; and with an air of conscious pride, Harold's guide conducted him to the cone of yellow, crinkled skin of which he was the lord and master. It was about the same size as an ordinary Indian wigwam, and put together in much the same way: long, thin pieces of driftwood formed the ribs, and a row of heavy stones was placed around the bottom fringe, to make it secure against the frequent assaults of the wind. Hardly had Harold peeped into the tent than he backed out again with a celerity which suggested that something had scared him. But it was not his heart that failed him. It was something much less poetic. It was his stomach. For there, right at his feet as he entered, on either side of the narrow doorway, was the Esquimau's larder—two great piles of seal meat and blubber, as repulsive-looking a sight as it is possible to conceive, and giving forth an odour that surely only a native, educated to it from childhood, could endure for one moment. Either the sight or the stench would have been quite enough for Harold, but together they were simply overwhelming; and turning away he hastened out of the encampment in the direction of the shore.

His guide hurried after him with a very puzzled expression on his fat, dirty face, and seemed quite relieved when Harold, having by a great effort overcome his internal dissensions, said, with a reassuring smile,—

"Oh, it's all right! Made me feel a little sea-sick, that's all."

The "Huskie," of course, could not understand the words; but the speaker's tone and expression made it clear to him that he was not offended, and that was

"Harold missed a stroke with the paddle."

sufficient. Harold continued his progress toward the shore, for he had noticed a couple of kayaks, or Esquimau boats, drawn up on the beach, and he was anxious to examine them closely. He had read much about these curious little boats, which are to the dusky dwellers in the region of eternal ice what the birch-bark canoe is to the red man of the forest. He had seen many pictures of them, and now he was to see the thing itself.

There were two kayaks drawn up side by side, and on Harold showing his interest in them, his new-found friend at once indicated by signs that the larger and better one of the two was his own property. By dint of much gesticulation, Harold made him understand that he would like to see how the skiff was managed; whereupon the "Huskie" ran back to his tent for his paddle, and then launching the kayak, got carefully in, and with sure, strong strokes, sent it leaping swiftly over the still surface of the harbour. He was evidently a most expert kayaker, and as the boy watched him darting hither and thither, spinning round with almost startling suddenness, and seeming more like some huge water-bird than an awkward bit of humanity, he thought he had never seen so delightful a method of navigation, and he was determined to try it for himself. Accordingly, he beckoned to the Esquimau to come in, and when he had done so, indicated to him by signs that he would like the loan of his kayak for a few minutes. The Esquimau demurred at first, whether because he was afraid of some accident happening to the kayak or the boy, it is impossible to say. But Harold was not to be denied. He had managed birch-bark canoes and wooden *Rob Roys* successfully at home, and he could surely get on all right enough with a kayak.

After much persuasion the Esquimau gave a reluctant consent; and throwing off his coat and boots, Harold, in great glee, prepared for his experiment. The kayak was launched again, and brought alongside a big stone, where the owner held it while Harold stepped in gingerly, and stowed his legs away under the tiny skin deck. He was then handed the paddle, and with a strong shove sent out into deep water. No sooner was he thus committed to his own resources than he began to regret his rashness; for of all the cranky crafts that he had ever tried, nothing compared with this kayak. It seemed to have but one object, and that was to upset. Only by the greatest care and constant use of the paddle as a balancing-pole could he keep right side up. His progress, it need hardly be said, was exceedingly slow. Yet he did succeed in getting about a hundred yards from the shore, when he thought he had better turn.

Captain Marling had by this time concluded his visit, and was returning to his gig, when he caught sight of Harold crawling tremblingly over the calm water in the kayak.

"Just look at that young rascal!" he exclaimed, turning to the factor. "That's just like him—always doing something rash. He's not afraid of anything."

"He's doing very well for a first attempt, as I suppose it is," said Mr. M'Tavish. "But see! he's trying to turn. I'm afraid he'll find it a ticklish operation."

"Be careful, there, Hal!" shouted the captain, in stentorian tones, as he saw the boy's danger.

But the warning came too late. In his efforts to right about face Harold unhappily missed a stroke with the paddle, and instantly his cranky craft capsized, taking its occupant down with it; for he could not disengage his legs from the narrow space in which he had been sitting. To the captain's horror, nothing was to be seen but the wet bottom of the kayak glittering in the sunshine.

CHAPTER VI.

ON THE TRACK OF HENRY HUDSON.

WITH a tremendous shout of "Man the boat! Get to your places!" Captain Marling bounded toward the gig, sprang in over the bow, and reaching the stern, grasped the rudder lines, while the men tumbled hastily into their seats; and the factor, putting forth all his strength, pushed the boat from the beach out into deep water.

"Now, then, pull for your lives!" cried the captain.

The men needed no urging. They loved their captain, who was as kind and just as he was strict; and they loved his son, whose bright, manly face was always welcome in their forecastle, when he came to listen to them spinning yarns, and very probably bringing with him some dainty from the captain's table. With stalwart strokes they sent the swift gig over the smooth water toward the spot where the curved bottom of the upturned kayak showed above the surface like the back of a porpoise at rest.

When first upset, the light skiff could be seen moving in a way that showed the boy imprisoned beneath was struggling hard to free himself. But ere the gig got more than half-way to it, the motion altogether ceased; and the captain noticing this, cried hoarsely to his rowers, "Pull, men, pull, or it will be all over with the boy!"

If it were possible to put forth any more strength than they were already doing, the men did it; and with six tremendous strokes sent the boat flying to the side of the upturned kayak.

"Easy all! Easy now!" cried Captain Marling. "Stand by to help!" And he sprang into the bow of the gig as it ran up alongside the kayak.

"Now, then, lift there." And he grasped one end of the light skiff, while a sailor seized the other. They quickly turned the treacherous craft over, to find poor Harold lying lifelessly underneath.

"God help us! Can he be dead?" cried the captain, in tones of agony, as he lifted the limp form of his boy into the boat, and pressed his hand against his face, which was ominously white and cold.

"He's just lost his breath.   He'll be all right, sir," said one of the men reassuringly.

"Pull for the ship as hard as you can!   Give way now with all your might!" the captain shouted, dropping down upon the stern sheets with Harold's body in his arms.   The men rowed their best, and the *Narwhal* was quickly reached—Harold being at once taken to the saloon, where he was stretched out upon the table, and restorative measures promptly applied under the direction of Dr. Linton.   For a time it seemed as though all efforts would be unavailing. The water into which he had been upset was deadly cold, and had almost instantly chilled him to the heart; and though he had been less than a minute actually immersed, the time was long enough to bring him very near to death.

Presently, however, under the skilful handling of the doctor, signs of life began to make themselves apparent.   A faint tinge came into the pale cheeks, and steadily deepened, the eyelids fluttered as though they were striving to open, and a gentle sigh was breathed from the lips that had seemed as though they would never move again.   At the end of an hour's unremitting effort he opened his eyes widely, looked about him at the group of anxious faces around as if he would like very much to know what all the fuss was about, and then fell asleep again.

"He's all right!" exclaimed Dr. Linton joyfully. "Danger's over. Let's put him snugly in his own bunk, and by to-morrow morning he'll be not a bit the worse for it."

"God be praised!" said the captain reverently. "I never knew how precious my boy was to me until now." And bending over Harold, the big man kissed him again and again with almost womanly tenderness.

One of the most intensely interested spectators of all these proceedings had been Patsy Kehoe. He knew nothing of Harold's mishap until his unconscious form was lifted gently on board; but when he saw his face of deathly paleness, and his utter helplessness, he sprang at once to the worst conclusion, and uttered an agonized cry of "Ochone! ochone! the dear young master; is he dead entirely? The Lord have mercy upon him!"

"Shut up, and be about your business!" growled old Strum, who was not disposed to look upon the stowaway with much favour. But Patsy was not to be sent away. With the tears running down his cheeks, he followed the sorrowful procession to the cabin, and there stood beside the table, swinging his hands and saying softly to himself, "The Lord help him," while his eyes watched everything that was being done, as though his own life depended upon the success of the means employed. When at length Harold breathed and opened his eyes, poor Patsy's glad relief could not be suppressed, and taking up the boy's hand, he covered it with tears and kisses, exclaiming joyfully, "Sure, he's not dead at all; indeed he's not. Oh, the dear young master!"

Even in the midst of his own deep emotions Captain Marling could not help observing Patsy's display of genuine feeling, and his heart was touched by it.

"That youngster's got something good in him," said he. "I'm glad now I let him stay with us."

The doctor's prophecy in regard to Harold proved correct. He spent the rest of that day in his berth, and was a trifle shaky the following morning, but otherwise was not a whit the worse for his very narrow escape from death. When he reappeared on deck, the *Narwhal* had left Nachvak far behind—Captain Marling having succeeded in securing a satisfactory interpreter—and was now at the entrance to Hudson Strait, where another hard struggle with the ice was anticipated.

At first the prospect for a successful run seemed

very good, the field ice being loose and rotten, and permitting the steamer to plough steadily through it at the rate of from six to eight knots an hour. But as the day wore on, the weather, unfortunately, set in thick, a dense fog hanging over the sea making it necessary to slacken speed, especially as the appearance of many "growlers"—fragments of large ice-bergs—hinted broadly at the proximity of big bergs, and made a sharp look-out imperative. Steaming either at half-speed or dead-slow, the *Narwhal* crept cautiously onward, the fog occasionally lifting and allowing a sight of land to be obtained. It was a very dreary business, particularly when, as happened more than once, it was found advisable to tie up to a huge floe, and resign all idea of progress for the time. One such floe was three hundred yards long by two hundred yards wide, and at least twelve feet thick—a perfect island of ice, which would have carried the *Narwhal*, solid and heavy as she was, on its broad, white back as easily as a feather's weight.

The two sealers whom he had taken on at St. John's being thoroughly acquainted with the locality, and the peculiar kind of navigation the steamer was now experiencing, Captain Marling put the ship in their charge, and allowed himself more leisure than he had done since the commencement of the voyage. This was pleasing to Harold, who was never so happy as when in his father's company, and the captain took the opportunity to tell him something about the famous bay toward which they were directing their course.

"If you could only see, my boy, the kind of ships in which the first Englishmen sailed across the ocean to discover this continent, you would just set them right down for lunatics. Why, the *Discoverie*, in which Henry Hudson had the pluck to push his way through all the ice and fog and other dangers of these straits into the heart of the big bay beyond, could easily stand on the *Narwhal's* deck, and yet leave us plenty of room to work the ship. She was a crazy little craft, not much better than one of those ballast hookers that we laugh at in Halifax harbour, and yet in her Henry Hudson, more than two hundred and fifty years ago, crossed the Atlantic, ventured through these straits, and made his way clear down to the southern end of the bay, where he stayed all winter. Poor fellow! he deserved a better fate than fell to his lot," added the captain musingly.

"Why, father, what happened to him? Was he frozen to death in the winter?" inquired Harold eagerly.

"Worse than that, Hal; worse than that," answered the captain. "Oh, how I'd like to have my hands on the scoundrels! Wouldn't I string them up to the yard-arm at short notice! You see, Hal, when they were on their way back the next summer, they had a mutiny on board the *Discoverie*, and the rest of the crew forced Captain Hudson, his son, and seven men who stood by him, to get into a boat with just a little water and food, and then the scoundrels cut them adrift and sailed away. Not one of them was ever seen again. Just fancy, Hal, if the *Narwhal's* crew were to play that game on us!"

"Not much fear of that, is there, father?" answered Harold, in a very confident tone, yet at the same time drawing nearer to his father and taking hold of one of his hands; for the thought of Henry Hudson with his young son adrift in the merciless ice in an open boat took hold of his quick imagination, and he could not restrain a shudder as he asked, "What do you think became of them, father?"

"I am sure I don't know, Hal," replied Captain Marling. "It was in mid-summer the villains cut them adrift, and the bay would be full of floating ice. Poor fellows! no doubt they did their best to make the land. But even if they succeeded, they wouldn't have been much better off; for there were no Esquimaux along the east coast, and their only chance would have been to go clear across the country to Nachvak— a matter of hundreds of miles, which of course was an utter impossibility."

The gallant discoverer of the great bay that now bears his name, and particularly the boy that shared his sad fate, were often in Harold's thoughts after this, although he little imagined that the time would come when the same kind of villany that had succeeded in Hudson's undoing would bring his father and himself into deadly peril.

On the fourth day the fog cleared, the sun shone out bright and warm, the ice to a large extent disappeared, and putting on a full head of steam, Captain Marling sent the *Narwhal* along at a fine rate of speed. The unclouded sky, the pleasant air, the rapid progress, brought back everybody's good-humour again, and hearts were light and countenances cheerful as the good steamer ploughed swiftly onward. To avoid the bulk of the ice which the current carries down from Jack Frost's fastnesses along the southern side of the strait and out into the Atlantic Ocean, there to resolve itself back into its original water, Captain Marling coasted the northern side of the strait, and

Harold had a fine opportunity of studying an Arctic landscape.   As they passed along, they saw huge beetling cliffs, broken here and there by stretches of low land, or rather rocks, the relics of last winter's snow lying in small patches in the gullies, while on the sunniest slopes of the hills a faint tinge of green could now and then be detected.   The picture was a very dreary one.

"I should die if I had to live in such a country as this," soliloquized Harold.   "No trees, no flowers, hardly any grass.   Why, what could a fellow do? He couldn't play cricket, or base ball, or tennis, or anything of that kind.   He couldn't go swimming, for the water's too cold.   I'm precious glad I don't live here, any way—aren't you, Patsy?" he added suddenly aloud, as he noticed Patsy coming toward him along the deck.

"What's that, Master Harold, if you please?" asked Patsy.

"Aren't you glad you don't live over on that land all the year round?"

"Troth, that I am, sir," replied Patsy, fervently. "It's little liking I have for Griffintown, but I'm thinking that same place is a mighty sight better than this awful hole.   Sure, my eyes are achin' for a bit of a tree, and there isn't so much as a leaf in the whole place.   But what's that, Master Harold?" he exclaimed excitedly, grasping Harold's arm and pointing to the wide stretch of open water on the seaward side of the steamer, where his keen eyes had detected a thin column of water, like a fountain in a flower-garden, rising from the sea.

Harold looked very hard, but being a little too late, saw nothing.

"There! there it is!" cried Patsy, pointing to the spot with a trembling finger, and this time Harold saw it plainly.

"It must be a whale!   Let's call father," exclaimed Harold, running to the head of the companion-way and shouting at the top of his voice, " Father, father, come up!   We see a whale."

Captain Marling was in the cabin consulting some charts, but at the first sound of his son's voice he threw them down and dashed up on deck with an agility that could hardly have been surpassed by Master Harold himself.   In the meantime the news had become known to all on board, and the seaward bulwark was crowded with faces eagerly scanning the rippled blue plain, from the midst of which the significant fountain had appeared.

"There she spouts!   There she spouts!" would be the cry, as the fountain played again.   The whale, indeed, seemed to be doing its best to attract attention. It was a whale of the right kind, of medium size, and was coming toward the ship in a slanting direction, in evident innocence of the presence of those whose special mission it was to wage war upon itself and its kindred. Happily, however, for this particular monster of the deep, and much to the chagrin of those on board the *Narwhal*, preparations had not yet been made to engage in its pursuit; and there was nothing to do but look idly on, and watch the great creature flinging its challenge into the air, and rolling its vast bulk over the waves, just as if it understood perfectly that the *Narwhal* was not yet ready to do it harm.

Frank Lewis was anxious to get out his rifle, and see if it were possible to kill the whale with a bullet, since the harpoons were not in readiness; but Captain Marling would not suffer it.   He was a tender man at heart, and he could not approve of what would be, after all, a bit of needless cruelty; for even though the whale were killed, it would be sure to sink before the gear could be rigged to deal with it properly.   So it was permitted to go on its way unscathed, the men watching it with longing eyes, as turning off at an angle, probably because it discovered the presence of the steamer, it went away down the strait, giving a farewell spout ere it finally vanished.

The boys naturally felt very proud of having sighted the first whale; and Captain Marling, by way of encouraging such sharpness of vision, promised them a sovereign gold piece for each time that either one of them would be the first to point out a "fish." For be it known that, although whales are not really fish at all, but warm-blooded mammals, the whalers will never call them anything but fish.

The appearance of the whale filled the *Narwhal* with excitement and bustle. The unexpected difficulties encountered on the passage northward had delayed Captain Marling very materially. But now he was determined to make up for lost time, and orders flew thick and fast as the day drew toward its close. In forty-eight hours they ought to be right in the midst of the whaling-ground.

CHAPTER VII.

PREPARING FOR ACTION.

WITH wind and weather favourable, and but little ice obstructing her progress, the *Narwhal* steamed on through Hudson Strait, then through Fisher Strait, and thence up into Rowe's Welcome, where Captain Marling proposed to make his first attempt against the poor whales, whose only crime was their possession of such valuable blubber and bone.

On board ship everything was at fever heat of activity. A great deal had to be done before whaling could be properly entered upon, and there was work for everybody. The crew of the *Narwhal*, including the captain, consisted of fifty men, some of whom bore very odd titles; for whalers seem to have taken a good many of their business terms from the Dutch. There was, for instance, the speksioneer, the officer under whose direction the whale is cut up; the skeeman, whose duty it was to superintend between decks the stowing away of the blubber in the tanks; and others with titles equally queer. The *Narwhal* had a fine outfit of eight whale-boats, and there were therefore eight harpooners, including the mates and the speksioneer, eight boat-steerers, including the skeeman and boatswain, and eight line-managers, the duty of the latter being to pull the stroke oar in the boat, and to see that the lines are coiled away clear, so that they will run out freely when a fish has been struck.

The first thing to be done was to get the boats out, they being always stowed away under deck until the whaling-grounds are reached. It did not take long to have them up on deck, and, after a minute examination at the hands of the captain and carpenter, thoroughly cleansed from all dust and dirt. They were beautiful boats—long, low, and narrow, sharp at both bow and stern, and painted pure white, with a broad crimson stripe a little below the gunwale. Each boat would carry six men, five to row and one to steer, the harpooner pulling the bow oar, and having command over all. The steering is done, not by a rudder, but by a long oar which projects out over the sharp stern, and with which the steerer can sweep his obedient boat round upon her track in one-half the space in which it could be done by a rudder.

All the boats having been gotten ready, the crew next turned their attention to "spanning on"—that is, attaching the lines to the harpoons and coiling them away in the boats. The bustle reached its height over the operation, for each boat's crew did their best to have their craft equipped first; and the rivalry and excitement were very keen, as under the captain's watchful and approving eye they toiled away like beavers. Harold had already attached himself to Frank Lewis's boat, and he got very much worked up over the contest, running about from boat to boat to see how the other competitors were getting on, and then back to his own to encourage its men by assuring them that they were bound to win.

This was what they had to do. Each boat carries two harpoons—a gun-harpoon and a hand-harpoon. The gun-harpoon is made wholly of iron, but the hand-harpoon has a long wooden handle that makes it look something like an old-time spear. To the harpoons is first fitted the "foregore" or "foregauger" —that is, a piece of white untarred two-and-a-quarter inch hemp rope from three to twelve fathoms long, which is much stronger and more yielding than any ordinary rope. Then to this foregore the remaining whale-lines, of which there are five to each boat, are carefully spliced—the result being a line more than six hundred fathoms in length, or a little over half a mile. A pretty long fishing-line; but then, of course, it is intended only for the biggest kind of fish, if it be right to call a whale a fish.

This mammoth fishing-line is then carefully flaked down in the stern sheets in a compartment made for the purpose, with the exception of about one hundred fathoms, which are flaked down in a box in the centre of the boat called the "foreline beck," and of the "foregore," which is coiled in a small tub or kid in the very bows of the boat, right alongside the gun. The proper disposition of the line is a matter of the highest importance; for if this should be carelessly done, and the line should happen to catch when the harpoon is fast in a whale and the monster is "sounding"—that is, diving into the depths of the ocean—there is no telling what the consequences might be; but the chances are ten to one that the big whale-boat would be instantly dragged under water as though it were a feather. Consequently, while the men, in their anxiety to be first at the finish, worked with all their speed, yet it could be easily seen that there was no actual haste. Nothing was slurred over, but everything was done thoroughly; for might not their own lives pay the penalty for slighted work? Frank Lewis aided his men by hand as well as by voice, and thus working away together they got the lead of all the other boats, Harold's shrill shout of " First—first! We've finished first!" presently announcing that the second mate's boat had won, whereat the others cheered heartily, and offered him their congratulations; for it is considered an omen of good-luck to have your boat ready first.

The lines having been satisfactorily stowed, the next proceeding was to prepare the whale-boat's armament, which is quite an extensive one. First, there were the two harpoons—the harpoon-gun, which is fixed on a swivel on the bows, so that it can be turned in any direction; and the hand-harpoon, that lies beside the gun, the handle resting on a " mik " or crutch, ready for immediate use. These harpoons are made of the softest Swedish iron, so that they may readily bend without snapping, and the distortions they sometimes undergo when in use are really wonderful. Each of the *Narwhal's* harpoons had the steamer's name plainly stamped upon the shank, so that if the whale should happen to get away, trailing the broken line after him, and they should happen to meet him again, there would be no difficulty in recognizing him. Besides the two harpoons, each boat carried four lances for killing the whale after it has been struck, a tail-knife used for cutting holes in the tail and fins of the dead whale, a hatchet for severing the line should that be necessary, a fog-horn for signalling in event of being caught in a fog, two boat-hooks, two small buckets for pouring water over the lines when they are running out very fast, to prevent their setting the boat on fire in consequence of excessive friction, and a number of other things that need not be detailed.

To stow all these things neatly and securely away in a small boat, bearing always in mind how absolutely important it was, not only that everything should have a place, but that it should stay in that place, and not get in the way of the lines, no matter how much knocking about the boat might have, was a task requiring skill as well as care, and the day was fast declining before all necessary arrangements were completed, and the last of the boats had been made ready and swung into its place at the davits.

"There, now," said Captain Marling, with a sigh of

satisfaction; "we've only got to put up the crow's nest, and we're ready for business."

The putting up of the crow's nest did not take long, it being simply a large cask which is triced up to the main-royal mast-head, the lower end resting on an iron jack, and the upper part being bound to the mast by an iron strap. In the bottom is a small trap-hatch, just large enough for a man to crawl through. He then shuts it down and stands upon it. As the crow's nest is a very exposed, cold place, it is lined with furs to protect its occupant; and so long as the vessel is on the fishing-ground, the cask has always somebody in it during the day, sweeping the sea with his telescope in search of signs of whales.

No sooner had the crow's nest been hung up on high than Harold was possessed with a wild ambition to get into it. As it was nearing dusk, Lewis tried to dissuade him, saying that it would be a better time in the morning. But Harold was not to be put off; so, his father making no objection, he proceeded to run up the rigging until he reached the cask, and then found little difficulty in shoving up the hatch and creeping through to the inside.

"Hurrah, boys! This is the place for me!" he cried triumphantly, as from his cozy citadel he looked down upon the deck beneath, where the men were still moving busily about. It was a very interesting picture that lay spread out before him. Right underneath was the noble steamer, cutting her way through the blue water, already settling down for its night's rest; and so smooth was her progress that Harold could faintly feel the throb of her engines, even up in his lofty eyrie. Far away to the right a dim, dark line showed where the land broke the otherwise clear horizon, and here and there level patches of white, or tall fantastic forms, indicated the presence of the ice, which is indeed never entirely absent from these quarters. So widely scattered was this ice as to offer no obstacle to the progress of the *Narwhal*, and at the rate the steamer was now going she ought to reach Rowe's Welcome by sunrise.

Seeing Patsy emerging from the companion-way, Harold hailed him with a loud "Hello, Patsy! how is this for high?" which made the boy look up in a puzzled way, for he had been below while the crow's nest was being put in position, and so knew nothing about it; but he had no difficulty in making out who called him, and shouted back, "Well done, Master Harold! What's the weather like up there?"

"Splendid!" replied Harold. "Won't you come up? Come along!"

Now Patsy was even more expert in the rigging than Harold, and it seemed but a few seconds before he reached the bottom of the cask and knocked for admittance. On attempting to join Harold inside, however, he found himself somewhat at a loss. Harold was standing upon the little trap-door, which could not therefore be opened, and there was not room enough in the crow's nest for him to stand aside and allow the trap to be pushed up. After consulting together a while, the boys hit upon a plan which successfully met the difficulty. Harold climbed up to the top of the cask and sat upon the edge, holding on by the tips of the mast, which projected a little distance beyond it, and then, the coast being clear, Patsy had no difficulty in effecting an entrance, so that in a moment or two the boys were standing together in the crow's nest.

It was, of course, pretty close quarters, but they did not mind that. They had the same delightful sensation as is experienced by travellers who have accomplished the ascent of some lofty mountain peak, and they promised themselves frequent trips to this exalted position. The sea had by this time fallen into almost complete calm, and although the dusk was drawing nearer, they could see a good distance from the steamer. They were, of course, sharply on the look-out for whales, and ere long their search was rewarded by the discovery of something large and black swimming rapidly in the direction of the *Narwhal*. Instantly they raised the cry of "A whale! a whale! We see a whale!" and Lewis, who was at this time walking the quarter-deck, it being evening watch, shouted back to them eagerly, "Where is he? point him out."

"There! there!" replied Harold, indicating the spot with his outstretched arm, while Lewis brought his glass to bear upon it. Lewis looked long and carefully, and then calling to him one of the men they had taken on board at St. John's, asked him to look. Then they had a short consultation together, the boys watching them impatiently, and wondering why orders were not given for the boats to be lowered right away. But evidently no such orders were to be given, for shaking their heads in a way that implied it was not necessary to do anything, the men parted, and Lewis called up to Harold, "Only a bottle-nose, Hal! Not worth going after."

"Only a bottle-nose?" exclaimed Harold, turning inquiringly to Patsy. "What does he mean?"

" Faith, I don't know," replied Patsy. " But I'm thinking that if I don't get back to my work the steward will be after me with a stick." And so saying, the active little chap dropped down through the hatch, and made his way to the deck with the ease of a monkey in his native woods.

Harold intended to follow him at once, but a movement on the water near the ship attracted his attention, and he lingered on until at length darkness closed around him, and Lewis shouting up to him, " Say, Hal, are you going to stay up there much longer ? Hadn't you better come down ? " he prepared to return to the deck.

No sooner had he started than he began to regret having postponed his descent so long. It was easy enough getting into the crow's nest in broad daylight, but it was a very different matter getting out of it again in the dark. Lifting up the hatch, he cautiously dropped his feet through, and felt around for the first rung of the ladder beneath. But somehow or other he seemed unable to find it. After several unavailing efforts, he drew himself back into the cask, feeling not a little nervous. His chief difficulty was that he had to hold the trap up while crawling through, and it was a pretty heavy affair, being almost the full size of the bottom of the cask, and strongly put together. He felt very much inclined to call to Lewis to come up and help him out of his difficulty. But against this his pride rebelled. Patsy had gotten down all right; why shouldn't he? He would try again. Moving very carefully, he once more dropped through, holding to the cask with his right hand, and holding up the trap with his left. After feeling around for a moment, and stretching his leg out as far as it could go, he succeeded in touching the ladder rung, and with a thrill of relief dropped upon it. But as ill-luck would have it, his foot slipped off the rung, the whole weight of his body came upon his right arm, and instinctively he let go the trap that was held up by his left, and grasped the bottom of the cask with both hands. The heavy hatch thus released fell at once, striking him a cruel blow upon the top of the head, and forcing his head down between his shoulders. The blow was very severe, and but for the soft, thick cap that partially protected his skull, it would certainly have stunned the boy. As it was, it made him feel giddy and faint, and only by that supreme effort which imminent danger can call forth was he able to retain his hold and save himself from being dashed to the deck. More than this he could not do, and in that perilous position he hung, realizing his danger, but powerless to rescue himself from it, while he heard, as in a dream, Lewis's voice calling up from below, " What are you about up there, Hal ? Going to stay all night ? Come down, or your father will think something's happened to you."

CHAPTER VIII.

AMONG THE MONSTERS.

HAROLD'S long delay in coming down from the crow's nest made Lewis feel somewhat anxious, and when there was no response to his call, he thought it best to go up at once and see what was the matter. Accordingly, he hastened up the rigging with the case of a well-trained sailor, and the first thing he knew, his face came in contact with a pair of heels that dangled downward in a very strange fashion.

"Good heavens! what's the meaning of this?" he exclaimed, feeling about to find out where the rest of Harold's body was. "What's the matter with you, Hal?" he cried out.

"I'm caught here, and I can't move," answered Harold faintly.

"Oh, is that all?" replied Lewis, in a tone of relief; "I'll soon let you loose. Just hold on a minute longer." Then looking downward into the abyss of darkness below, he shouted, "Ahoy! on deck there!"

" Ay, ay, sir!" answered one of the watch.

" Just bring me up a lantern here, and look sharp about it," commanded Lewis.

The man hastened to obey, and in a few moments the welcome light came swinging up the mast. It at once made the situation clear, and while the sailor held it, Lewis had no difficulty in extricating Harold from his perilous position; and as the boy was already recovering from the effects of the blow, the descent to the deck was easily accomplished.

" That was a funny sort of a scrape for you to get into," said the second mate, when they were on firm planks again. " How on earth did you manage it?"

Harold described just how it happened.

" Humph!" said Lewis. " We must have that hatch fixed so that it will stay up when anybody's climbing in or getting out. We can't have it playing pranks like that."

Captain Marling looked a little grave at first when Harold told him of his adventure; but when, after a careful examination of the boy's head by the surgeon, no further injury than a big bump was discovered, and Harold assured him that he was really all right now, his countenance cleared as he said,—

"Oh, well, you can't learn to be a sailor, and especially a whaler, without taking plenty of hard knocks, my boy, so we won't say anything more about this one. And now to bed, Hal; there'll be plenty to do and see to-morrow, if fortune favours us."

The prospects were bright when Captain Marling came on deck next morning, and he looked the very picture of good-humour as he moved briskly about, making sure that everything was in perfect readiness to go in chase of the first cetacean that might come into sight. The *Narwhal* was moving at a moderate speed through the wide waters of Rowe's Welcome. The day was clear, and not too warm. The wind played gently over the blue plain, which was dotted here and there with floes and fragments of ice that offered no obstacle whatever to the steamer's advance. A more auspicious day could not have been desired, and every heart on board beat high with hope. Aloft in the crow's nest one of the Newfoundlanders, glass in hand, scanned the sea carefully, while those on deck looked up at him every moment, impatiently awaiting the signal that their prey was in sight.

In this way the morning hours passed, and still the

occupant of the crow's nest kept silence, until those below began to grow weary and restless. The engines were no longer used, as the slightest noise made by the screw, or, in fact, by anything *under* water, would be sure to scare the wary and watchful whale, were he in the neighbourhood. But with plenty of canvas set, the *Narwhal* bowled along, tacking hither and thither, so as to cover as much of the fishing-ground as possible.

At length, as noon drew near, the long-awaited and welcome cry of " A fish ! a fish !" came down from the look-out ; and instantly all was excitement on board. The maintop-sail was backed, and two boats quickly lowered and sent in chase. The second mate's was one of the two, and Harold begged very hard to be allowed to go in it. But his father thought he had better stay on board; so he had to be content to watch the proceedings from the cross-tree, into which he climbed as soon as the boats left the ship.

Pulling with all their might, the sinewy oarsmen sent their graceful boats leaping over the waves toward the whale, which was in full sight not more than a quarter of a mile distant from the ship. Before they were half-way to it the monster sank, causing Harold to utter a groan of disappointment. But the men in the boats knew better; and spreading out so as to cover plenty of ground, they rowed on more slowly, awaiting the whale's reappearance. The minutes passed while everybody fairly trembled with excitement; and then, at last, the black rounded mass of the whale's back was seen to rise close to the boat on which Frank Lewis was bow oarsman and harpooner. The others held their breath as the handsome second mate laid his oar fore and aft, and rose to handle his gun. Quick, yet cautious, was every movement. The great musket was pointed at its mighty target. For an instant Lewis glanced along the barrel; then there came a flash, a puff of smoke, a moment of intense anxiety, followed by a joyful cry from the crow's nest of " A fall! a fall!" that was echoed from the successful boat, making it beyond doubt that Frank Lewis was fast to a fish.

Immediately the steamer's deck was alive with men, all frantically shouting, " A fall! a fall!"—that being the whaler's term to designate a harpooner's success in striking his huge game—and rushing to the boats, to make ready for instant departure.

Captain Marling ordered four more boats to be lowered, instructing them to spread out in different directions, so that some one of them might be near the whale when it rose to blow, it having, as usual, "sounded" the moment it received the keen harpoon. The rapidity with which his orders were carried out spoke volumes for the discipline prevailing on board his ship. Each man knew his boat and his place in that boat. There was the liveliest possible bustle, but no confusion. Splash! splash! splash! splash! and one after another the four graceful boats were dropped from the davits.

"Stand to your oars! Give way!" cried their captain. And off they went in the direction of Lewis's boat, upon which the "jack"—the flag—was now flying, in token of his success. Spreading out widely, their harpooners kept a diligent look-out for the monster below, while the boat that had started at the same time as the second mate's drew up alongside of her, in order to bend on its lines, should that be necessary.

"There she blows!" cried the men in one of the boats presently. And, sure enough, the whale rose suddenly to the surface, not more than a hundred yards away, and sent up into the air a fountain of water, whose reddish tinge showed that Lewis's harpoon had reached a vital part; and then, no doubt becoming aware of the presence of the boats, it dived into the depths again, the lines running out at a rate that made the bollard head smoke.

"Confound her! she's making for the ice!" exclaimed Lewis, in tones of mingled apprehension and irritation, as the movement of the lines showed that, after sinking to a certain depth, the stricken monster had turned in the direction of an immense ice-pack that lay to the northward, and was hastening toward this shelter at railway speed. A turn was now taken in the line, so that instead of running out it held fast, towing the boat along, with her bow almost on a level with the water. Hatchet in hand, the second mate stood ready to sever the tightly-stranded rope the instant it threatened to drag the boat under altogether. The other boats followed as best they could, and presently they were all brought to a stop by reaching the edge of the pack, under which the whale had dived, intending, no doubt, to come up again in the first water space it could discover.

As luck would have it, the breeze that had been slight all the morning now freshened considerably, causing the heavy floes which formed the outer edge of the pack to tumble about in a way that was full of danger to the frail boats, which would be crushed like egg-shells were the ice to nip them. So severely were they being handled by the floes, some of which were drawing over twenty feet of water, that grave fears were entertained for their safety, and on more than one occasion they were only saved from being crushed by the prompt action of the crew, who, jumping hastily out, would haul them up on the ice.

Observing their danger, Captain Marling having already steam up on the *Narwhal*, pushed her into the midst of ·the loose heaving pack, and with no little difficulty picked up the boats, one after the other, the last being the boat that was fast to the fish, from which, by dint of considerable trouble—for the whale was still taking the line—the line was transferred to the steamer. No less than ten lines, or twelve hundred fathoms—that is, about a mile and a quarter of rope—had been run out by this time, and the captain began to feel anxious, lest, owing to the bothersome ice, not only the prize but a good part of these precious lines would be lost also.

However, it was no time for despair, and hoping for the best, all hands manned the line, even Harold and Patsy lending their young strength, determined to bring the fish home, pull away the line, or draw the harpoon. Everybody did his best, for was there

not a prize worth a thousand pounds at least at the
other end of the line ? and had not the captain promised to share the proceeds of the voyage in an unusually liberal manner among the crew, since not one of
them was to be omitted in the division of the spoils ?

. Then came a marvellous exhibition of the enormous power of this leviathan of the deep. That it
should tow a light whale-boat along without diminishing any of its speed was nothing wonderful ; but
now it was fast to a powerful steamer of full six
hundred tons, weighted with engines and heavy stores,
and yet so vast was its strength that this huge mass
was towed at a rapid rate through the pack, colliding
with and bumping against the great floes that continually obstructed its progress.

In spite of the constant danger from the ice, the
prospect began to brighten. Although a tremendous
strain had been brought to bear upon the line, both
it and the harpoon held, more than half of it had
been hauled in, and the men were singing cheerfully,
and already counting up their money. The captain's
face had lost its anxious expression, and beamed with
hope instead, while Frank Lewis's shone with pride
because of his good fortune in being the first to get
a fish. Hand over hand, yard by yard, the tough

line came dripping on board, as the big vessel bumped through the ice in the wake of her strange tug-boat. Three-fourths of it at least was now coiled up safely on deck, and the strain upon it seemed to be slackening somewhat, as though the whale was beginning to tire of its extraordinary exertions, when there came a sudden pause, then a mighty rush; the line ran out again at such a rate, that, thinking to check it, Lewis ordered a turn to be taken around the capstan; the steamer's speed greatly increased, when all at once a great floe came right across her path. It was too wide to avoid. There was no alternative but to charge straight upon it. With a crash that shook her stout frame from stem to stern, and sent Harold, who had been standing in the bow, an intensely interested spectator, headlong to the deck—raising a laugh among the sailors, for he was not hurt a bit— the *Narwhal* struck the floe, crushed through it nearly one-half her length, and then stopped dead. Instantly the line strained as tight as a fiddle-string.

"Ease it! ease it, for your lives!" shouted Lewis, who at once saw the danger. But it was too late. There was a quick, sudden jerk, a sharp snap. Amid deep growls of disappointment the line fell slack, running in easily as the men once more began to haul upon it. The harpoon had drawn, and after four hours of plucky fighting against its human foes, the mighty cetacean had come off victorious, bearing with it a deep and painful wound in token of the conflict.

The disappointment on board the *Narwhal* was very keen. From Captain Marling down to Patsy Kehoe, no one attempted to disguise his feelings. The lost whale was of the largest size, and would have been a very valuable prize; and they had worked so hard for it. But it was "no use crying over spilt milk." The whale was gone, in all probability never to be seen again. Yet there was consolation in the fact that it had not taken any of the lines with it, although it had bent and twisted the harpoon so as to render it entirely useless.

Late in the afternoon the steamer managed to extricate herself from the ice-pack which had so provokingly interfered with her pursuit of the whale, and being once more put under canvas, made her way into a regular bight formed by the ice, where she "lay to," as it seemed a very promising fishing-ground. In both cabin and forecastle the talk that evening was of nothing but whale. In the latter place, the men were disposed to take a gloomy view of things, considering it ill-luck to lose their first fish; but in the cabin Captain Marling's habitual cheeriness had by this time reasserted itself, and the prospect was discussed with very great composure. Harold, of course, listened to everything that was said with open-mouthed interest, asking a question whenever he could get a chance, which, however, was not often, for the talk was very brisk; while Patsy made every possible excuse for lingering in the cabin, in order that he might lose as little as possible of what was being said. The Newfoundlanders, both of whom had been to Rowe's Welcome and other northern fishing-grounds on previous occasions, told some thrilling tales of their experiences with whales. According to their own accounts, they had been the victims of about every mishap that could happen to a whaler. Their boats had been smashed into splinters by a stroke of the monster's tail, or drawn right under water through the line failing to run out fast enough, or been crushed between colliding floes; and yet they had come off scathless and able to boast of at least a dozen whales apiece as trophies of their skill with the harpoon.

The more Harold heard of this the more it fired his ambition to go out in one of the boats the next time a whale was sighted, and he pleaded so hard for permission, that when the second mate came to his support, and promised to be personally responsible for him if allowed to go in his boat, Captain Marling at last gave way, and made the eager boy's heart jump for joy by saying, " Very well, Lewis; you may take him with you. You see that he keeps out of mischief."

CHAPTER IX.

A GOOD DAY'S WORK.

DURING the next week whales were seen constantly, and the deck of the *Narwhal* was a scene of bustle and excitement from daybreak till dark, the whole fleet of eight boats being at times sent out in pursuit of the great creatures that showed themselves in the distance. But so far success had held aloof. The whales seemed unusually wary, and after a hard pull of two or three miles a boat would get perhaps almost within striking distance, only to have its intended victim make off under the ice, where pursuit was impossible. All sailors are more or less superstitious, and if there is one class of them more superstitious than any other, it is the whaler.

Accordingly, they set themselves diligently to seek out the cause of their ill-luck, and resorted to many absurd expedients in their endeavours to avert the evil chance. As one day followed another without bringing better fortune, they began to show signs of depression, causing Captain Marling to feel as anxious to break this spell of failure for the sake of his men as for his own sake.

At length the tide turned. Pushing her way through the ice, the *Narwhal* reached a wide bay or inlet which was entirely clear of pack or floe, and here the captain felt sure there would be fish to take. Two boats were ordered to keep " on bran "—that is, manned and ready to start off at a moment's notice—the steamer being brought to a stand-still, and the boats drifting about in her immediate neighbourhood. Captain Marling himself spent much of his time in the crow's nest, Harold often being with him, as he swept the circle of his vision with his glass, alert to discover the first sign of the presence of whales. One fine, clear morning, when father and son were thus employed, the captain happened to be looking due north, while Harold was looking due south, when Harold suddenly grasped his arm, and pulling him around, pointed away off, about three miles distant, exclaiming breathlessly, " There, father ! what's that ? "

Captain Marling brought his glass to bear upon the spot to which Harold pointed, and the moment he looked his face became radiant, and turning toward the boats "on bran," he shouted joyfully, "A fish! a fish! a whole school of them, right off to the south! Make ready every boat on board'"

The excitement that followed was indescribable. Every man on board the *Narwhal* was in motion, and to an unpractised eye it might have seemed a scene of hopeless confusion. But this was not the case at all. Each man knew his place and his work. The confused crowd soon resolved itself into groups gathered about each boat, and with a quickness hardly credible the boats were dropped into the water, the men into their seats, and with a hearty "Good-luck to you, my boys," they were off to meet the approaching whales.

In the stern of the second mate's boat sat Harold, scarcely able to keep his seat for the fever of excitement that possessed him. The instant his father had confirmed his hope that the black spots far to the south were possible prey, he had slipped out of the crow's nest, and not missing his footing this time, scuttled down to the deck, where he posted himself in close proximity to Frank Lewis, determined that the boat should not go off without him.

The whales were coming directly towards the ship, there being at least a dozen of the monsters, ploughing their way through the water rippled by a gentle breeze, and sending up little fountains from the blowholes in their mighty heads. As they would soon become aware of the steamer's presence, and in all probability take to the depths at once, it was necessary to meet them before they came too near, and accordingly the men bent to their oars with an energy that called for every ounce of muscle in their sturdy frames. Yet not a word was spoken. Neither did the oars, vigorously as they were being pulled, make the slightest sound beyond a faint splashing, for thick thrum mats lay on the gunwale between the thole pins. These precautions were necessary because of the exceeding quick hearing of the whales.

Dividing into two groups of four each, the boats spread out so as to allow the procession of whales to pass between them, thus affording the best possible opportunities for attack, and thus they shot over the waves toward their gigantic prey.

Lewis's boat led the group to the right, and Harold felt as though he could hardly breathe as they drew nearer and nearer to the big fish without their presence being observed by them. The steersman stood high in his place, one hand grasping the great oar, to every movement of which the swift boat responded

like a thing of life, and the other keeping time to
the strokes of the rowers. A few minutes more, and
at a signal from him Lewis drew in his oar, laid it
carefully fore and aft, then turned his face toward
the bow, and put his hands upon the harpoon-gun.
They were now almost within striking distance. A
few strokes more, and the critical moment would
come. The oarsmen strained yet a little harder; the
boat fairly leaped over the water. The second mate
having given one sharp glance along the line to make
sure that all was clear, pointed the gun at the unconscious whale: there was an instant's pause, then a
loud report, and then the glad cry of "A fall! a
fall!" rang out over the waves, for Frank Lewis's
luck had once more asserted itself, and he was fast
to a fine big whale.

But the luck was not with him only. Scarcely
had his shout of triumph reached the ship, when it
was followed by the same inspiring sound from others
of the boats, until, before the school of terrified
whales sank out of sight, no less than four boats were
fast, and the men of the *Narwhal* were in for a most
exciting and exhausting struggle. This time the fates
seemed altogether favourable. The bay was entirely
clear of ice; only a gentle breeze stirred the surface

of the blue waters. The whale-hunters had the whole day before them, and no reason appeared why they should not make prizes of all four of the huge creatures in which their harpoons were now fastened. Two of the fish were rather small, but the other two —one of them being Frank Lewis's—were of great size, and worth perhaps almost two thousand pounds apiece.

The instant the whale to which the second mate was fast felt the keen harpoon boring into its vitals, it threw up its tail and dived into the depths, the line running out at such a rate that the smoke arose in clouds from the bollard head, half shrouding Lewis, who, lance in hand, stood up in the bow, ready to give his captive its *coup de grace* so soon as he could get near enough. The men drew in their oars, and the light boat fled like a shadow over the waters, as the stricken monster vainly sought safety in flight. So desperate were its efforts to free itself, that nearly a mile of line was taken before the running out ceased. Harold's heart was palpitating with delight. He thought that in all his life he had never experienced anything half so glorious as this, being towed along at the speed of an express train by a giant fish. Not a mite of fear or nervous-

"*Lewis stood up in the bow, ready to give his captive its coup-de-grace.*"

ness had he. He had no time for that. The struggle was too absorbing to allow him to think of anything else.

"My, but this is grand!" he exclaimed to the steersman. "I never had so much fun in my life."

The steersman gave him a pleasant smile, to show that he heard him, but was too engrossed in his work to make any reply; which, however, made little difference to Harold.

"Take in line," called out Lewis; and with the speed and skill that come only from long practice, the men drew in the dripping line, and coiled it away in its nest, ready to run out again should it be required. Fathom after fathom it came swiftly in, until at least one-half of it had been recovered.

"Stand by now, and be ready to give way," was Lewis's next order; and the men put their oars in position, while all waited with bated breath for the reappearance of the whale, which must soon come to the surface to breathe. One, two, three minutes passed, and then suddenly, so close to the boat that the commotion it caused set it to rocking, the huge black, glistening back of the whale rose out of the water, and a stream of water deeply dyed with blood shot up in the air.

"There she blows! there she blows! Give away on your left there, hard!" shouted Lewis, grasping the long, keen lance in his right hand; and round swept the boat in the direction he desired. "Now, then, all together," he cried again. The men bent to their oars. The stout ash blades bent beneath the strain upon them, and the boat almost sprang out of the water in response. Three mighty strokes, and Lewis was close beside the whale's forefin. For an instant the polished lance flashed in the sunlight, then it sank up to its haft in the soft flesh.

"Back water, for your lives!" shouted the second mate, dropping down into the bottom of the boat, and the oarsmen sent the boat backward. They did so just in time. Maddened by this fresh attack, the whale lifted its terrible tail high in the air, and brought it down with awful force upon the very spot where the boat had been the moment before. The spray from the blow drenched every one on board, and the boat rocked as though in a whirlpool.

"No, you don't," cried Lewis joyously. "That was a close shave. But a miss is as good as a mile."

As if disgusted at its failure to crush its tormentors, the whale sent up a spout that was nearly all blood this time, and then "sounded" once more.

But evidently its end was near. The line did not run out at all so fast as before, and only a few hundred yards had been taken ere the creature returned to the surface a short distance from the boat.

"We've got our fish right enough this time," said Lewis, smiling broadly. "We'll just stand off until she 'kicks the bucket.'"

For a few minutes the whale lay still upon the water, as though resting, and the tired men were glad to rest also. Then came the final flurry. The huge frame trembled all over, the deadly tail was lifted and brought down upon the water with re-sounding blows, spout after spout of dark heart's blood incarnadined the sea, and then all was still.

"Pull up now, men; it's all over," ordered Lewis, after waiting a little while to make sure it was the case.

The boat drew alongside of the mighty carcass, a hole was cut in each fin, the fin tow passed through them, and the big fins lashed tight to the sides, so as to offer no obstruction to towing. The tail was then secured to the stern of the boat, and the prize in this manner towed to the steamer, which, happily, was not far away.

When they got time to look about them, those on the second mate's boat saw with delight that one of the other boats had already killed its fish, and was making toward the ship, while the other two were still fast, with good prospects of like success.

The towing of such an unwieldy prize was no easy task, but they had only half a mile to go, and their hearts were light; so with cheery songs they tugged away, and in due time were alongside of the *Narwhal.* Approaching at the port side, the fish was brought between the fore and main rigging, and made fast by a rope around its tail that passed through a block on the fore-mast, and another rope through a hole in the under jaw that was rove through a tackle on the main-mast. The whale was on its back, and the right fin, which was next the ship, was dragged taut up, and secured by a chain to the upper deck. A stout wire rope, stretching from the main-mast to the fore-mast, and known as the "blubber guy," held four large blocks, through which were rove the fore and main "spek" tackles, whose use was for hoisting on board the huge layers of blubber, some of which would weigh between one and two tons. The "kent" or cant tackle having been rigged, the object of which is to turn the fish over as it- is being flinched, everything was in readiness for this interesting operation. The men were duly refreshed from the ship's stores, and then the work of flinching the whale began.

Harold looked forward to this with intense curiosity, and posted himself in his favourite eyrie in the main cross-tree, where he could overlook everything without getting in anybody's way. He did not expect it to be as interesting as the chase of the whale, but it could hardly help being well worth seeing; and such, indeed, it proved to be.

In the port main rigging was the captain, superintending the whole business; at the gangway stood Peter Strum, with eye watchful to see that every command of the captain was minutely carried out. Upon the upturned belly of the whale jumped the eight harpooners, their boots being armed with iron spikes, to prevent their slipping, and at it they went with their keen blubber spades and knives.

First of all, a strip of blubber, nearly a yard in width, was cut from the neck, just abaft the fin; and a large hole being cut in the end, the strap of the cant tackle was passed through it, and by this means the fish could be turned over as desired. With spade and knife the men cut big strips of blubber from the belly, which were one by one hoisted on board the steamer, where they were received by the boat-steerers, who with long knives cut them into pieces about two feet square, and passed these pieces over to the line-managers, whose work was to seize them with pick haaks, or pickies, as they are called for short, and send them shooting through a small hole in the main hatchway to the deck below. Here they were taken charge of by the "skeeman," and by another man, oddly denominated as the "king," and stowed away between decks until a favourable opportunity should come for the final operation of "making off."

When the blubber had all been removed, the precious whalebone, worth between two and three thousand pounds a ton, and of which the whale would yield a good part of a ton by itself, was carefully detached from the vast mouth, and lifted on board by special tackle. Then the great tail was cut off for a purpose that will be afterwards explained; and thus stripped of everything of value, the "kreng" or carcass was released, disappearing with a plunge into the green water, which it turned to blood for some distance, while the men sent up a lusty cheer by way of a farewell.

The men had worked hard and well, only two hours being required to dispose of the first whale, and were allowed a brief rest and another "lunch all around" before attacking the second. For the *Narwhal* had been wonderfully fortunate, three out of the four whales having been secured—the two big fellows and one of the small ones; and the day's work would "pan out" at least six thousand pounds: so that it is safe to say that from Captain Marling down there was not a merrier—nor a dirtier—crew afloat than toiled and laughed and joked and shouted on board the steamer all through that long mid-summer afternoon.

Harold found abundant amusement for a long time in watching all the bustle and noise, and then after it became somewhat monotonous he discovered another way of entertaining himself, which proved so diverting that he felt bound to call Patsy up to share it with him. No sooner had the process of flinching begun than the steamer was surrounded by hundreds of fulmar petrels, or "mollies," as the whalers call them—noisy, greedy, quarrelsome birds, in appearance much resembling the ordinary sea-gull, that clamoured and fought over the numerous pieces of kreng and blubber wherewith the water about the ship was liberally sprinkled. So fearless did their insatiable voracity render them, that they would even alight upon the whale within reach of the men, who would often catch one of them and fling it back into the midst of the flock swimming eagerly alongside, producing a disturbance that very soon subsided. They were not worth killing, and Captain Marling would not suffer Harold to shoot them; but they were a great nuisance, so he had no objection to his making a mark of them in throwing. Accordingly, the two lads brought up a bucket of small lumps of coal, and had fine fun seeing which could make the better shots, using the coal for ammunition. The "mollies" were pretty cute, and could dodge the missiles with good success; but their voracity often betrayed them, for they would run the risk of being hit rather than lose some choice morsel upon which they had decided, so that as between hitting and missing, honours were about even.

Just before darkness infolded the ship in its soft embrace for the night, the work of flinching was finished, and to the accompaniment of a lusty cheer that actually terrified the mollies, and went echoing out over the still water, the third and last kreng sank into the depths, while the tired men, all smeared with blood and blubber, indulged in a grand clear up before tumbling into their berths.

Harold soon followed their example, and his dreams were full of whales and mollies and exciting incidents, but they contained no prophecy of what awaited him on the morrow, and in blissful ignorance of coming peril he slept as only a weary boy can sleep.

CHAPTER X.

IN PERIL.

THE deck of the *Narwhal* presented a most unattractive sight when Harold stepped out upon it the following morning. As the process of "making off" was yet to be performed, no attempt had been made to clean it up, and it was in an indescribably filthy condition, and so slippery that the most experienced sailor had to ponder the path of his feet pretty carefully if he did not want a roll into the lee scuppers. After the carpenter, however, had sprinkled sawdust over the worst places, it was not so bad, although Harold could not help regretting that it was necessary to make such a mess of the steamer that had always hitherto been kept as neat as Aunt Etter's best parlour.

It was a fine calm morning, and there were no whales in sight; so Captain Marling ordered all hands to turn to and assist in the business of "making off."

This was not so interesting as the work of flinching. The great strips of blubber, weighing from half a ton up to two tons apiece, were hoisted up on the main deck, and there divided into pieces about twelve or sixteen pounds weight by men called "krengers." These pieces were then passed on to the "skinners," who removed the tough, strong skin, and handed them over to the choppers, who, using big chunks from the whale's tail as chopping-blocks, chopped the large pieces up into little ones, and these, by means of a canvas shoot called a "lull," were sent down the hatchway into the tanks waiting below to receive them, where they would remain until despatched to the refinery at the end of the voyage.

It was an animated and cheerful sight, even if not a particularly picturesque one, that the main deck presented during the "making off." The precious blubber, looking more like huge lumps of cheese than anything else, came tumbling up out of the hold, and went from krenger to skinner, and from skinner to chopper, growing smaller and smaller in its journey, until at last, reduced almost to mince-meat, it disappeared down the hull, and vanished into the dark iron tanks, to be disturbed no more for perhaps many months.

Harold, of course, must needs be krenger, skinner, and chopper by turns ; and although he was inevitably more of a bother than a help, and gruff old Peter Strum would have banished him to the stern if he could have had his own way, Captain Marling thought it best that his son should know the business of whaling from start to finish, and accordingly the boy was given full liberty to do what he pleased in the matter. It was little short of a miracle that he did not cut off some of his fingers in his frantic efforts to vie with the expert choppers, and he did manage to give himself a cut in his boyish zeal. But it was only a slight one which a bit of sticking plaster made all right.

From morning till noon the men toiled away like beavers, and were just resuming after having stowed away a good dinner, when the cry of " Whale, ho ! whale, ho ! " from the crow's nest, which all this time had never been unoccupied, threw them into a state of excitement and confusion. Dropping their knives and choppers, they swarmed to the steamer's side, and there, sure enough, not a mile away, two whales could be seen spouting and rolling about on the water.

Captain Marling was at first quite in a quandary.

There was still sufficient work to occupy his crew for the rest of the afternoon, and the "making off" must be finished now that it was begun. One whale on board was worth two in the water, and the blubber could not be neglected. On the other hand, the whales just sighted were evidently fine large ones, and it seemed too bad to let them go without having a try at them.

At this juncture the second mate presented himself, and, touching his cap respectfully, said,—

"I should like to have a go at those whales, sir. Do you think you could spare my boat?"

The captain hesitated a moment. Lewis's first whale had by this time been safely stowed away in the tanks, and he was therefore the best entitled to be sent off again. "It seems a pity to lose such a good chance, sir," urged Lewis.

The captain pulled out his watch, glanced up at the sky, which was beginning to be clouded over as though thick weather was at hand, rubbed his chin thoughtfully, and at length said,—

"Very well, Lewis. But I can't spare more than your crew. So you'll have to try it alone."

"All right, sir," replied Lewis, his face lighting up with joy; and then he shouted,—

"Crew number one get ready to launch at once."

The men singled themselves out from the crowd, and while their shipmates looked enviously upon them, threw down the tools with which they had been working, and hastened to launch the boat. Within a few minutes after Lewis's order the boat was off, and rowing swiftly in the direction of the whales, which had changed their course since first sighted, and were now making toward the mouth of the bight in which the *Narwhal* was anchored, they having perhaps in some way scented the presence of danger.

Captain Marling's attention was called off by something just as the boat was starting, and he did not notice that Harold, taking it for granted he could accompany the second mate on his chase, had leaped into the stern, where he sat looking very important when his father, catching sight of him, exclaimed,—

"Bless the boy! I didn't know he was going, and I don't want him to go, either."

His first impulse was to order the boat back and make Harold get out. But by this time it was full one hundred yards away, and to return would have caused the loss of too much time; so saying to himself, "Well, boys will be boys, and that boy is a chip of the old block, sure enough," the captain let the boat go on. Many a time during the next few days did he bitterly condemn himself for not carrying out his first impulse, and vow that could he have had a hint of what was to happen he would not have let even Lewis go. But beyond the slightly threatening aspect of the heavens, there was no cause for apprehension; so comforting himself with the thought that Lewis would be all the more careful because of having Hal with him, Captain Marling mounted to the crow's nest, whence he could watch the course of the boat as it sped on after its prey.

The whales were now quite a mile and a half distant, still keeping close together, and not going very fast. With strong, steady strokes, the oarsmen pulled their stout blades through the water, and the boat responded like a thing of life. The steamer rapidly receded, and the big fish drew nearer as the minutes passed. Not a word was spoken. A wave of the hand from the steersman was the rowers' guide, and presently at a gesture from him Lewis softly drew in his oar, and rising from his seat took his position on the bow behind the harpoon-gun.

They were now within a hundred yards of the larger of the two whales, and the great creature having just sent a stream of water whizzing into the air, was lying quietly upon the surface, as if taking a rest. Unfortunately, it was not in a favourable position for attack, the tail being turned toward its pursuers, and it was necessary to make a kind of detour in order to come at it on the right side. This was done with the utmost care, and in almost perfect silence; notwithstanding which, just as Lewis was levelling the gun at the monster, it took the alarm, threw up its tail, and vanished in a cloud of foam. In the hope of striking somewhere, the second mate pulled the trigger; but the whale was too quick for him, and the result was a clean miss.

"Too bad!" exclaimed Lewis. "Another second, and we would have had her. Lay to, there, and look out for her when she comes up again."

No one watched more closely for the cetacean's re-appearance than Harold, and as luck would have it, he was the first to discover it.

"There she blows!" he exclaimed, pointing about a quarter of a mile away to the left. And he had made no mistake. The whale rose, spouted, and then lay still again, with its full broadside presented to the boat.

"Hurrah! now's our chance. Give way for all you're worth," said Lewis; and away darted the boat.

This time there was no disappointment. The boat shot up to within twenty yards of the fish before its presence was suspected. Indeed, the first intimation the latter had of its proximity was when the gun went off with a loud report, and the keen harpoon clove its way deep into its vitals.

"A fall! a fall!" shouted Lewis triumphantly, the men taking up and repeating vigorously the joyful words. "Stand by to see that the lines run out clear."

Away dashed the wounded whale with tremendous speed, taking line at such a rate that the bollard head smoked as though it were on fire, and the end must soon be reached. Lewis's face grew grave as the line continued to run out and showed no signs of slackening. There were but a couple of hundred yards left in the coils, and hatchet in hand he was standing at the bow ready to sever the swift-running rope the moment the strain became great enough to drag the boat under, when happily the line slackened, no more ran out, and with a huge sigh of relief Lewis put down the hatchet, saying, "She's got enough. Stand by to take in as fast as you can."

The men laid hold of the line and drew it in hand over hand, while Lewis watched eagerly for the whale

to return to the surface. This it did nearly a quarter
of a mile away, and after taking breath "sounded"
again, but went so slowly—for it was evidently hard
hit in the lungs—that it was not necessary to let any
more line out; so taking a turn with it around the
bollard head in such a way that it could be instantly
loosened if required, Lewis allowed the whale to take
the boat in tow, feeling sure that it would soon tire
of the work.

In this, however, he was mistaken. The whale,
sore wounded as it was, did not soon tire. On the
contrary, it kept right on for mile after mile, steering
straight out into the open water, until the *Narwhal's*
hull began to sink below the horizon; and Harold,
glancing back over the course, could not help wishing
that he was on board the good steamer again, instead
of being far away from her in a small boat, following
a maddened monster into unknown perils.

They were at least five miles from the ship, whose
heavy spars already looked like delicate sprays of
fern as they outlined themselves against the horizon,
before their captive, or rather captor, paused in its
career and once more rose to the surface for breath.
The watchful Lewis saw his opportunity. The boat
was rowed swiftly to the creature's side, and a long

lance driven deep home to its heart ere it could escape.

"Back water, all!" he shouted. The obedient boat retreated rapidly. The dying whale lashed the water into bloody foam in its final flurry, and then lay still.

"We've got her!" exclaimed the men joyfully. And so they had. The whale, a fine large one, worth a pot of money, was dead beyond a doubt, and it now only remained to tow their prize back to the steamer. This they at once proceeded to do. The lines were gathered in and coiled in their compartments, the fins were well pierced and lashed under the belly, and the big tail was triced up to the stern sheets.

"We've got a long pull ahead of us, my boys, and I don't quite like the look of the weather. But we'll try it, any way," said Lewis, scrutinizing the sky in the direction of the *Narwhal* carefully.

The men were willing enough. They were working for themselves as well as for the ship. To bring the whale alongside safely meant a good many pounds for each, so they buckled to their task with hearty vigour.

But they were never to be any richer by this

capture. The fates had otherwise decreed. Just as they were getting well under way, the dark, threatening look of the heavens was explained. A sudden gust of wind, heavily laden with snow, burst upon them. So dense was it that they could not see more than fifty yards ahead, and so blinding and stinging that no one could attempt to face it. Hoping that it might prove only a brief squall, Lewis ordered the men to lay to on their oars, and wait until it passed over. But instead of abating, the storm continued, and seemed, moreover, to increase in violence. It soon became clear that the whale must be given up. To hold on to it any longer was simply to court death; so Lewis reluctantly gave orders to cut it loose, and presently the precious prize for which they had toiled so hard was drifting away, its life having been taken in vain.

The question, now was, how best to act under the circumstances. The wind, unhappily, was blowing straight from the direction in which they wanted to go, and so strongly that the boat could not face it. Their only safety lay in fleeing before the storm until its violence should be over. To enable them the better to do this, the short mast, which every well-equipped whale-boat carries, was set up, and the sail well reefed down, carefully raised and secured. Under this canvas, slight as it was, the boat fairly flew over the water, and being as seaworthy as a lifeboat, there was no more danger of its being swamped. But where were they going, and what was to become of them? Serious questions these indeed, and so all on board the flying boat felt them to be, as with anxious, frightened faces they cowered in their seats, dearly wishing they were well out of their predicament.

It was growing late in the afternoon. The dusk would soon be upon them, and then their position would be still more perilous. There was no longer any necessity for Lewis being in the bow, so he came down to the stern, where Harold, crouched in a corner, was doing his best to protect himself from the storm.

"Keep up a brave heart, Hal," said Lewis, putting his arm around him. "We'll come out of this scrape all right; never fear."

Harold lifted his head and looked into Lewis's face. It was clear that he had been crying; but at the sound of his friend's voice he wiped away his tears, and made a big effort to control himself.

"Will we be all right soon?" he asked, with trembling lips.

"Perhaps not very soon," said Lewis, hugging him close, for he was exceedingly fond of the boy. "But it'll be sure to be clear by morning, and then the *Narwhal* will come along and pick us up."

"I hope so. Dear father! how anxious he'll be! He'll come right after us, I'm sure."

"Of course he will. He's probably searching for us now, and maybe will pick us up before the morning," said Lewis, assuming his most cheerful tone. "But look here, my boy: you must have some better protection than that reefer of yours, or you'll soon be chilled through. Ah, I have the idea!" he added, springing up. Taking the tarpaulins that are used to cover the whale-lines, he made Harold curl up in the coziest corner of the stern, and then covered him completely, making him so comfortable that the tired boy, comforting his heart with Lewis's assurance of ultimate deliverance, soon fell asleep, and dreamed that he was safe on board the *Narwhal* again.

On sped the stanch boat through the darkness, Lewis taking turns at the steering oar with the steersman, and the men sitting still and silent in their seats, as though they had already resigned themselves to death. Toward midnight the snow gradually abated, and finally ceased altogether; but the wind held on, and constant care was required to keep the boat true in her course.

"The captain will have a tough job finding us if we go on all night at this rate," said Lewis to the steersman.

"Ay, indeed, sir; but I've been in just as bad a box before, and got out of it all right. If to-morrow's fine, our chances are good," was the cheering reply.

About this time Harold awoke, shivering with the cold and aching from his hard couch.

"Are we near the steamer?" was his first question, as he peered eagerly into Lewis's face.

How glad would Lewis have been to give him an affirmative answer! But there was nothing to gain by deceit, so he replied, "Not yet, Hal; we can't find the steamer until morning, you know."

"And when will it be morning?" asked Harold tremulously.

"Not for some time yet, Hal. But keep up your heart. The snow's all over, and the wind's going down; there's nothing to be afraid of," returned Lewis.

Brave words these, no doubt, and well meant, but they could not disguise from Harold any more than they could from Lewis himself the gravity of their situation. They must now be not less than fifty miles from the steamer, and increasing the distance all the time. Did they but know in what direction to steer, they might make some progress back over the course they had come, but to do this when so completely in the dark would be perhaps only to lessen the chances of their rescue. Then at any time the wind might change, the ice bear down upon them, and their good boat be crushed like an egg-shell between the pitiless floes. All on board realized the magnitude of the danger, but few spoke, and then only in subdued tones, as though little hope were left. And the air grew keen and penetrating, chilling the stoutest to the heart as the frail craft, with its crew of seven precious souls, sped before the wind none knew whither or to what fate.

CHAPTER XI.

IN TIME—THANK GOD!

MORNING broke clear and bright upon the lonely boat, with its chilled and weary occupants, and as their eyes eagerly swept the horizon, no ground for hope appeared to cheer and comfort them beyond the assurance that somewhere far up to the northward the *Narwhal* was already setting forth to their rescue. But how was she to find them? They made but a tiny speck on the vast wilderness of sea and ice about them. The little union-jack, which every English whale-boat carries, running it up to the mast-head whenever a fish has been struck, seemed like the mere mockery of a signal of distress, as they reversed it and drew it to the peak again. Then, even though the steamer should find the boat ultimately, might they not be dead from hunger and thirst before that would happen?

Just here they had good cause to be grateful that Captain Marling was a man of ideas, and that, having always the courage of his convictions, he never hesitated about putting his ideas into execution. Now, one of his ideas was that no whale-boat should leave the steamer without being provided against the contingency of separation from the vessel for perhaps a day or more. Not a season passed upon the great cod-fishing banks without some ghastly tale of fishermen out in their dories losing their schooner and drifting about to a slow death of torture from thirst and hunger. It were quite as likely for the same thing to happen in whale-fishing. Accordingly, the thoughtful captain had put to good use the compartment at the whale-boat's stern, by having made for it two strong tin cases, one of which was always kept full of pilot bread, and the other of water.

"We've got enough to keep us going for a couple of days, at all events, if we're not too greedy," said Lewis, as he proceeded to inspect the cases. They were both full to the brim, and held ten pounds of biscuits and two gallons of water respectively. Not much for six men and a big boy to live upon for any length of time, but not to be despised, you may be sure. They must be husbanded with greatest care;

for who could tell when the lost ones would be safe on board the *Narwhal* again ?

The sun rose in unclouded splendour, and its warm rays were unspeakably grateful to the shivering men. It was now the beginning of August, the hottest month of the brief Arctic summer, and the chances were that the storm which had come so inopportunely would be succeeded by a long spell of fine weather. They had therefore little to fear from the elements. Poor Harold, who had looked wofully haggard and forlorn at daybreak—for, if the truth be told, his tears had fallen freely through the night—grew much more cheerful under the brightening influence of the sun, and after eating a biscuit and taking a cup of water, felt a good deal more composed.

" It's going to be a fine day.   Father will be sure to find us, won't he ? " said he, looking eagerly into the second mate's face for confirmation of his hopes.

" Of course he will," replied Lewis, with a positiveness which thoroughly satisfied his questioner.   " If we're not all safe on board the steamer by night, I'll never prophesy again, that's all."

Lewis was entirely sincere in his answer.   He really thought the *Narwhal* would find them before night, little imagining how poor a prophet events would prove him.

On looking about them, it seemed clear enough that the storm had blown the boat across the open water, and that the vast field of ice to the westward, hilled and hummocked into all sorts of fantastic shapes, and stretching away as far as eye could see, was really shore ice, leading up to the solid land.

Landing upon this snow-white shore, Lewis climbed one of the highest hummocks in the neighbourhood, and swept the horizon with his keen glance. He had more than one object in view. Looking shoreward, he sought for some indication of the presence of Esquimaux, with whom as a last resort they might find a shelter; and turning seaward, he tried to discover the line of smoke that would be the first glad intimation of the steamer's approach.

But in neither quest was he successful. From horizon to horizon, north, south, east, and west, not another sign of life broke the dread monotony of ice and sea. The wilderness of water stretched along beside the wilderness of ice, and the keenest sight was powerless to find a bound to either.

"We may as well haul the boat ashore," said Lewis. "We'll have more room to move about on the ice."

So the boat was drawn up at a good place, and everybody landed, feeling very glad to be released from their cramped quarters. Had they been simply on a sealing expedition, for instance, they would have thought it pleasant enough ; for the air was warm, the ice firm and fairly level, and it would have been easy to have a lively time. Even as it was, the men's spirits manifestly rose, and they indulged in some sky-larking, which showed that they were by no means over-despondent.

Leaving them in charge of the boat, Lewis took Harold with him, and went off for a walk shoreward, wishing to ascertain, if he could, the breadth of the band of ice upon which they had landed. The walking was pretty difficult, and they could make only slow progress ; but they had any amount of time at their disposal, and the chief thing was to find occupation. It kept them from thinking too much. They had gone about a couple of miles, when Lewis's sharp eye noticed something moving in the snow several hundred yards distant, and he determined to investigate. Bidding Harold keep well behind him, he proceeded to stalk the game with the caution of an Indian hunter, making use of every hummock and inequality of the ice that offered concealment, Harold faithfully imitating his movements. Lewis had with him the hand-harpoon, a very formidable weapon, and also a good revolver in his hip-pocket, while Harold carried one of the lances; so that they were quite ready for any ordinary encounter.

Creeping forward noiselessly, they drew near the spot where the moving object had first been discovered. It probably had seen them, for it had slipped behind a hummock, and for aught they knew was making off on the other side. At length they reached this very hummock. Lewis motioned Harold to stay still, and then, throbbing with excitement, moved inch by inch around the base of the hummock, holding the revolver in his right hand and the harpoon in his left. Breathlessly Harold watched him until he disappeared round the corner, and then, unable to restrain himself, took a few steps forward. As he did so, he heard a shout from Lewis, followed quickly by the sharp crack of his revolver, and with a wild, startled roar a big white bear sprang out from behind the hummock and shambled off shoreward. The gait of a polar bear is an awkward enough movement under any circumstances, but there was something peculiarly clumsy about this one's mode of progression; and after watching it closely for a moment, Lewis shouted, in a triumphant tone,—

"Hurrah! his shoulder's broken!" This was precisely what bothered Bruin, and after getting over about a hundred yards of rough ice, he stopped, growled fiercely, and turned to face his pursuers. It was evident that he found locomotion altogether too painful to seek safety in flight, and had made up his mind to fight it out.

"Be careful now, Hal. Keep well behind me," said Lewis, advancing cautiously. Harold needed no such admonition. He had never seen a polar bear before, not even in a menagerie, and this great clumsy creature, with his long vicious head, huge paws, and shaggy fur of a dirty white tint, was not just the kind of new acquaintance he felt any desire to cordially embrace.

Covering the bear with his revolver, Lewis advanced until within safe shooting distance, and then, taking aim at his head, fired. The sharp crack of the revolver split the still air, there was an awful roar of pain and fury, and with a mad effort to reach his enemy the bear rolled over on the ice, to all appearance as dead as a door-nail.

Dropping his revolver into his pocket, Lewis

grasped the harpoon, and rushed forward to examine his prize. In doing this he showed a lack of his customary prudence which came near having serious consequences. A polar bear has a hold upon life that is almost beyond belief. If an ordinary cat has nine lives, this monarch of the North must have at least eighteen. Both of Lewis's bullets had taken effect—the first having broken the bear's shoulder, and the second having entered his skull, stunning him at once. Yet he was far from being dead; and just as Lewis ran up to him he suddenly sprang to his feet, reared up on his hind legs, and with a clever blow of his uninjured paw sent the harpoon spinning out of the second mate's hand, following the blow with so swift a rush that ere Lewis could evade him he had thrown himself upon him, and man and brute went down together in a heap upon the ice. For an instant Harold was petrified with horror. But it was only for an instant; then the boy's brave heart responded to the call of danger, and grasping the lance firmly in both hands he sprang forward, and the bear just then happening to lift his head, he thrust the point with perfect aim right into the monster's eye. Without a groan the bear rolled over, this time dead beyond a peradventure.

Too full of anxiety for his companion to feel exultant over the rare good fortune of his stroke, Harold dragged the bear off Lewis, and his heart almost stopped beating when he found him senseless. But feeling sure that he could not be really dead, he lifted his head into his lap and chafed it with his hands, crying frantically, "Mr. Lewis! Mr. Lewis! what is the matter? Oh, do speak to me!"

For a few moments he was kept in harrowing suspense, and then, to his unspeakable delight, Lewis opened his eyes, put his hand up to his head, and asked, in a faint, bewildered way, "What's happened? What's the matter?"

The reaction was too much for Harold's overstrained nerves, and bursting into tears, he exclaimed, "It's all right. He's dead. I killed him."

Lewis soon recovered his senses, and sitting up, looked from the dead bear to the weeping boy with an expression in which amazement and delight were oddly mingled.

"Did you really kill him, Hal?" he asked at length.

"I did indeed," replied Harold, wiping away his tears. "I jabbed him in the eye with the lance."

Satisfied that all danger was over, Lewis proceeded

to examine himself, and to his great relief found that beyond the bump from the ice, which had rendered him insensible, he had suffered no injuries whatever.

"I am not hurt a bit," said he. "If I'm not the luckiest fellow alive! Come, now, let us go back to the boat and get some help. We can't manage that great carcass between us."

The men of the boat were immensely surprised to hear of the killing of the bear, and taking a big piece of whale-line they dragged the body back to the boat, where they skinned it with their sheath-knives, and stretching out the pelt to dry, cached or stored the carcass in a hummock; for although polar-bear steak is far from toothsome, especially when raw, still it was better than no food, and there was no telling to what extremities they might be driven ere the *Narwhal* would find them.

On the particulars of the struggle being told, Harold became quite a hero in the sailors' estimation, and altogether this incident served to make a welcome break in the long day of waiting which passed without any sign of the steamer's approach. Once more the night closed in upon them, and with such composure as they could muster they prepared to make the best of their sorry plight.

The following morning found them beginning to show very plainly the effects of their continued exposure; for although the night was comparatively mild, and happily free from wind, it was, of course, impossible to obtain any real comfort, and sleep came only because of utter weariness. Poor Harold seemed so miserable in both mind and body that Lewis was filled with anxiety concerning him. The boy had a constitution of no ordinary sturdiness, but it was hardly calculated to endure long protracted exposure on an ice-floe. As the sun rose bright and strong in the heavens, a new danger confronted them. The white glare attacked their eyes, and they were threatened with snow blindness. This meant a serious addition to the perils already surrounding them, and the second mate warned his men to keep their eyes turned seaward, where the cool, pleasant blue-green of the water would do them no harm.

The long day dragged itself past with intolerable slowness. Owing to this new danger, none now dared to leave the boat, lest they should wander away and be lost. The pangs of hunger increased hourly, for the biscuits were well-nigh gone, and Lewis doled them out in little morsels that hardly made a mouthful. Their thirst they could partially assuage by sucking snowballs; but their only resource against starvation presently would be the raw, repulsive flesh of the polar bear. Sad-eyed and sick at heart, and yet marvellously patient, bearing their suffering with a stolid courage that filled Harold with wondering admiration and helped him to be brave also, the little group of castaways was once more enshrouded by the darkness of a night that for aught they knew might have no waking.

In the meantime what had they been doing on board the *Narwhal?* Every movement of Lewis's boat had been closely followed by Captain Marling from the crow's nest, and when, seeing the jack run up on the whale-boat's mast in token of a whale being struck, he joyously shouted, "A fall! a fall!" the busy men below paused in their work for a moment to give a rousing cheer at Lewis's success. So intent was the captain in watching the boat, that he did not notice the storm rushing out of the north until it was fairly upon him. But in an instant he realized the danger. Hastening to the deck, he ordered steam up at once. What yet remained of the "making off" must be laid aside, and all energies directed toward getting the steamer under way.

There was much to be done. The anchors had to

be raised—a slow process at the best of times—the dull-burning fires set going fiercely, the deck cleared of its many encumbrances, and all this took time; so that, despite the captain's impatient orders and the hearty vigour with which his men obeyed them, it was a full hour ere the *Narwhal* moved slowly out of the bight, and then with gathering speed set off in search of the endangered boat.

By this time the snow-storm was raging wildly, and it was impossible to make out anything fifty yards from the steamer.

"God help them, and take care of my boy!" murmured the captain, his heart aching with apprehension. Every man on board had but one thought—the rescue of their shipmates; and all who had no other duty posted themselves in the rigging or at the bows, heeding not the storm in their eagerness to descry some signs of the whale-boat. When night came on, the largest lanterns on board were hung at the mast-heads, and the steam-whistle was sounded at frequent intervals in hopes of attracting the lost ones. But the day dawned and brought no cheering news to the captain, who had not left the deck nor closed his eyes all night. "We'll be sure to find them to-day, won't we, Peter?" said he, with assumed cheerfulness, to the faithful old mate standing near him on the bridge.

"Ay, ay, sir; no doubt of it. They've just run off before the wind. We'll pick them up ere nightfall," replied Peter, in a tone of confidence that did the captain good.

But that day passed in fruitless search, and another night came, and the darkness that enwrapped the steamer was not deeper than the gloom which filled the captain's heart and affected every one on board. It was a sorrowful ship's company. Harold and the second mate were both prime favourites, the crew of the boat comprised the pick of the forecastle, and their absence made a gap which all felt acutely. Patsy Kehoe was almost beside himself with grief and anxiety. His warm, passionate Irish nature had gone out to Harold in a love that daily acts of kindness had formed and fostered until it became the very centre of his life. He scarcely slept or ate as the time passed and the missing ones were still unfound, and somehow Captain Marling came to feel as though there was no one who sympathized with him so fully as did the little Irish stowaway.

The third day came, and the weather was still fine. Having struck straight across the open water in the track of the storm until he reached the ice, the *Narwhal* had then coasted along its edge, the captain feeling sure that if anywhere the boat must be upon this ice. Unfortunately, he turned southward first, and thus lost a day, not turning north again until it was clear he had gone much further than the boat could possibly have done. Then he steered toward the north, and pushing the steamer to her utmost speed, scanned every foot of the shore ice as he passed.

The third day was drawing to its close, and, utterly worn out, the heart-broken captain had thrown himself upon the bench beside the cabin skylight and fallen into a profound slumber, when Peter Strum, then up in the crow's nest, shouted out at the top of his gruff voice, " Boat, ahoy—on the lee side ! "

Patsy just at that moment appeared on deck, and instantly catching the mate's meaning, he sprang to the captain's side, and shaking his arm vigorously, cried out in his ear, " Wake up, sir! wake up! Mr. Strum's found them."

Half stupified with sleep, the captain staggered to his feet, and looked about him in a bewildered way, and just then Strum's voice roared out again, " *Narwhal's* boat, and all hands safe."

With a fervent exclamation of "Thank God! thank God!" Captain Marling rushed to the rigging, and climbing up to within a few feet of the crow's nest, cried, "Where? where?   Show me where!"

There was the boat in full view, lying close to the edge of the ice, and there, standing beside it and making frantic signals to the steamer, were one, two, three, four, five, six men and a smaller figure that could not be mistaken.

On sped the steamer, straight toward them.   Right up to the edge of the ice she glided.   With the speed of thought a boat was lowered, and— Well, what more is there to tell, save that in a few moments Harold, pale, haggard, and weak, but living, was folded to his father's heart.

CHAPTER XII.

IN QUIET WATERS.

IT is easier to imagine than to describe the scene of gladness and gratitude which followed the reunion on the ice. Weak and faint as the rescued ones felt, the appearance of the *Narwhal* had infused new life into them, and they vied with their rescuers in the vigour of their manifestations of joy. The parties made a very lively group, and no one was more demonstrative than Captain Marling, who, having first assured himself that Harold had suffered nothing more from hunger and exhaustion than what a little care would soon make all right again, went from one to the other of the boat's crew, shaking their hands and clapping them on the back, giving full vent to his feelings. Then bethinking himself of their famished condition, he shouted out, "All hands on board to dinner;" whereat the whole party hastened back to the steamer as fast as hungry men could go.

A few days of rest and quiet completely removed all traces of their hard experiences from those who had been in such danger. Even Harold regained his plumpness and colour with a rapidity little short of surprising; and the interrupted process of "making off" having been duly completed, the *Narwhal's* prow was once more pointed northward.

It was now the middle of the Arctic summer, and the weather was almost uniformly fine and very delightful. Large quantities of ice continually appeared, but, as a rule, in a much broken-up condition, so that the steamer had little difficulty in forcing her way through it, although now and then the heavy floes would close in about her, and with every sail set and the engines under full steam, she would go boring and pushing her way through them into open water again. Had a gale sprung up while this was being done, the captain's skill would have been taxed to save the vessel from injury; for solid and heavy as she was, she would at times collide with an unusually big floe with such violence as to recoil several yards before again gathering way and charging her obstinate opponent. These collisions sometimes produced very ludicrous consequences—the sharp shock tumbling over members of the ship's company who were unprepared for such a sudden stoppage. Poor Patsy was one of the worst sufferers; for once, while in the act of carrying a tureen of soup to the dinner-table, he was sent headlong, the tureen being dashed to fragments at his feet, and he himself half drowned in its hot contents. Fortunately, the soup was not hot enough to scald him; but it would be hard to conceive a more comical object than he presented when he got on his feet again, with the thick, greasy soup spattered all over his face and chest. He was greeted with a roar of laughter as he stumbled into the saloon; and Captain Marling, enjoying the joke as keenly as anybody, hailed him with a hearty "Look here, Patsy; that's not the proper way to bring soup to the table. Go back and get some more, sir, and bring it to us properly,"—which order Patsy at once proceeded to execute.

They were discussing their future programme at the table that day, and the captain for the first time revealed a project that he had cherished from the outset, but which he had not intended to follow out unless he should be favoured with good fortune in his whale-fishing. This project was nothing less than to push on from Rowe's Welcome through Lyon Inlet into the wide waters of Fox Channel, hunt whales there until the tanks were well filled, and then, keeping still northward past the Melville Peninsula, venture through Fury and Hecla Strait into the great Gulf of Boothia, seeking a cozy corner of Committee Bay to winter in; and the following spring going on still northward to Lancaster Sound, thence eastward to the Baffin Sea, and, turning southwards, steer through Davis Strait into the broad Atlantic, and so home again.

It was a daring scheme, and as full of fascination for men of spirit as it was full of thought for men of reflection. Both classes were well represented in the little group that listened to the captain unfolding it; and if at first, with the exception of imperturbable old Peter Strum, their breath was taken away by its boldness, they soon regained their composure, and settled down to discuss the matter calmly, Harold listening with all his might, and hoping no less intently that his father would have his own way.

The pros and cons were simply these:—No one on board had ever been in those ice-bound waters before, the best charts of them were little more than guess-work, and the prospect of a long, dark Arctic winter away in the heart of that mysterious region could hardly be called a pleasing one; so that, aside from the ordinary perils of northern navigation—such as being nipped by the floes, or smashed into by an iceberg, or wrecked by a storm—there were good grounds for hesitation.

On the other hand, there was much to be said in support of the scheme. Gallant old Commander Baffin, in a small and crazy vessel of only fifty-five tons, had penetrated successfully the then undiscovered regions, part of which now appropriately perpetuates his name; and if his poor ship was equal to the ordeal, how much more the strong and sturdy *Narwhal*, more than ten times her size! In every man who is a man the spirit of adventure and discovery is strong, and here was a famous chance to give it free exercise. The route Captain Marling proposed to take was comparatively unknown. Few ships, and certainly no whalers, had ever been over it. Who could say what discoveries might not be made by enterprising men having every possible advantage, when such wonders had been accomplished by others not one-half so well equipped? There were provisions on board for two full years at least, fuel in abundance, furs in plenty for everybody—nothing lacking, in fact, that Arctic explorers would need.

All this Captain Marling laid before his listeners

with an earnestness that made him positively elo-
quent, and it was not long ere Frank Lewis became
as enthusiastic as himself. Dr. Linton was the next
convert; and then, more slowly, the two engineers.
As for the old mate, what the captain thought he
thought, and that was an end of the matter for him.
There were, however, two members of the group that
had been gathered together for consultation who very
clearly did not yield to the captain's persuasive elo-
quence. These were the two sealers shipped at St.
John's, Newfoundland, by name Joseph Collins and
Lemuel Stacey. They were by no means attractive
men, and more than once Captain Marling regretted
having taken them on board; for although their experience in northern navigation was very valuable
at times, still, upon the whole, their assistance was
never absolutely indispensable. And they were both
such rough, sullen, selfish men that their presence
was a positive blight, and everybody always felt
much relieved when they took themselves off in the
evenings to their own state-room; which, fortunately,
was as a rule quite early.

They had listened to the discussion without taking
part, not speaking until appealed to by the captain,
when Collins, in his grim way, urged a number of

objections against the scheme; and finding them, one by one, met and overcome by the others, relapsed into a sulky silence, from which he refused to be drawn again. Stacey, on his part, had nothing to say at all, but evidently gave tacit support to his countryman. Yet their silence by no means meant assent; and as will be seen farther on, their opposition, instead of dying away, became of increasing strength, and eventually bore very troublesome fruit.

Day succeeded day of sunshine and warmth, and they were easy times on board the *Narwhal*. The fires were put out in order to save coal, and under a fair press of sail the steamer bowled pleasantly along, tacking hither and thither according as the ever-present ice required or as the captain might direct. He was in no hurry. He would have liked very much to come across a couple more whales, and the crow's nest was never without its sharp-eyed watcher. But he knew very well that August was usually a blank month for the whalers. Just why it was so none could tell. That it was so, all could testify from actual experience; and so he was not disappointed at the time passing without any whales making their appearance.

There was one desire the captain was anxious to

gratify, and that was to do some sealing before the winter set in. The best time for that had of course already passed ; but away up in these vast solitudes, where they were rarely if ever disturbed, they must still be found in plenty, and to secure a few thousands of their pelts would be a very good way of filling in the month. At Repulse Bay the captain had his wish.

The *Narwhal* had been brought to anchor there for a day or two, and became surrounded at night by an immense ice-pack, upon which, when morning broke, seals were discovered in great numbers. Immediately the steamer was thrown into the liveliest excitement. Not a man stopped to think of breakfast, but, snatching up a ship's biscuit, crammed it into his mouth or his pocket, and grasping a handspike, an iron belaying-pin, or whatever else would serve as a club, hurried on to the ice in pursuit of the seals.

Of these there were many thousands scattered over a great field of fairly level ice, which had, by the force of the wind, been consolidated, so that there were few open spaces, and consequently there was no danger in going about. Captain Marling was one of the first upon the ice, Harold following close in his wake, and they were soon in the midst of the slaughter. It was indeed a slaughter. The poor seals made hardly any attempt to resist or escape, although here and there a big bull would bravely show fight, only to be knocked over ignominiously for his pains. The clubs were plied vigorously by the powerful sailors—the two Newfoundlanders, Collins and Stacey, being particularly active, and looking positively happy for once, as with tremendous blows they knocked their prey over on right and left.

Harold soon got very sick of the business. The seals seemed so helpless, and there was something wonderfully appealing and pathetic about the expression of the smaller ones, as, apparently recognizing the futility of attempting to escape, they lay panting on the ice, and looked up into the faces of their pursuers.

"Come along, Hal," cried his father, noticing that the boy hung back, and did not seem inclined to follow the retreating seals. "You've got a good club. Don't waste any time. There'll soon be no seals left."

"No, father, I think not. I don't like this work; it seems too cruel," replied Harold respectfully.

"All right, my boy," answered the captain, appreciating Harold's motives; "I won't urge you. Do just as you please." And then, as he hastened off after the others, he added, "Don't go far away; the ice may change at any time."

"I'll take good care, father," said Harold; and turning off to the right, he went over to the edge of the ice-field, where the open water was. He had not been there long before an incident occurred that made the operation of seal-killing more distasteful than ever.

The poor seal has a hard time of it. In the water the sharks and the sword-fish pursue him with unappeasable maws; and when to escape from them he betakes himself to the ice, the polar bear, Esquimaux, and sealers give him no peace. Even the sun becomes his enemy by raising blisters on his back, that cause him intense suffering, and make him dread returning to the water; so much so, indeed, that if found by sealers in that condition and pushed off the ice into the water, he will scramble back on to the floe at their very feet, seeming to prefer death by their clubs to further suffering.

It happened that several of these strange sea monsters called sword-fish were in waiting at the edge of the ice. No doubt they had followed the seals thither, and driven them up on the ice. The sealers were now acting in their favour by driving their prey back to them again. As Harold stood near the jagged edge of the ice-floe, a little band of seals, fleeing from their human enemies, scuttled swiftly past him and plunged into the water, that looked so like a haven of security. But, alas for them, it was nothing of the kind. Hardly had the green water closed over them than they reappeared at the surface, barking and lashing the water as if in a state of frenzy.

At first Harold, watching them with breathless interest, could not make out the cause of this disturbance; but presently it became clear enough. The sword-fish were in the midst of the helpless seals, which, barking, splashing, diving, sought in every way to evade their greedy enemies. The commotion was tremendous, and soon the crimson hue of the water told that more than one seal had fallen victims to the cruel sword. Thrilled with excitement, wishing with all his heart that he could take the part of the seals, but, of course, utterly powerless to do so, Harold watched this one-sided struggle going on. And then a very remarkable thing happened. The seals were now climbing back upon the ice, quite regardless of the boy's presence, of which, indeed, they seemed quite unconscious. About ten yards away Harold noticed a small seal swimming straight toward him with all its might, and not far behind it a huge sword-fish, evidently in hot pursuit. The seal, fortunately for it, had a head start, which would just about enable it to reach the ice; and this it succeeded in doing the instant before the sword-fish, coming on with a terrible rush, furious with disappointment, dashed its mighty weapon into the ice at the very spot where its intended victim had escaped. The ice, thick as it was, trembled with the force of the blow; and the terrified seal, uttering most piteous cries, hurried toward Harold, who was standing spell-bound at this wonderful sight, and, before he could move, placed its shiny head between his two knees in unmistakable appeal.

Harold was so astonished that he hardly knew what to do, but mechanically stooped down and patted the head of the seal, just as if it were a dog. The poor hunted creature looked up at him with eyes of wonderful softness and beauty; and there and then Harold determined within himself that if it were possible to get the seal safely to the steamer, he would adopt it as his pet, and perhaps take it back to Halifax with him.

Sitting down on the ice, he took the creature's soft, smooth head, all dripping as it was, between his hands, and fondled it tenderly. Strange to say, it made no resistance whatever. On the contrary, it seemed to thoroughly trust the kindness of its new-found friend, and its pantings and moanings ceased altogether as it lay restfully at his feet. Harold was full of delight. He had often read of seals making capital pets. In fact, there had been two in a pond in the public gardens at home for some time that he had frequently visited, and had seen them allow their keeper to play with them as if they were puppies. He was not long in making up his mind, therefore, to adopt this seal that had come to him so strangely as his pet.

But how was he to get it to the *Narwhal?* That was a problem which had to be solved first, and it did not look like an altogether easy one. To carry the creature was out of the question; it was quite too heavy, even if it did not object to being treated like a baby. To lead it was no less impracticable. There seemed but one way of meeting the case, and that was to drive the seal. Harold held a light club in his hand, and getting behind his voluntary captive, he waved it over its back, at the same time saying, in encouraging tones, "Get up now, Shiney. Off you go. Make for the steamer as fast as you can."

At first the seal went very well, and by shaking the club to right or left, as the case required, Harold was able to make it keep a pretty straight course for its destination. But after going half the distance it got tired, and seemed to think that it was already quite far enough from the water; for, in spite of the threatening club, it turned around and refused to advance an inch, sending forth a pitiful moaning that was almost human in its expression.

Harold did his best to coax it along, saying reassuringly, "Come along, Shiney. I won't hurt you; I'll be so good to you. Oh, come along, won't you?"

But Shiney was proof against all his blandishments; and in his despair he was about to try to pick the ungrateful animal up, when Lewis came along, and seeing Harold's predicament, called out, "Hello, Hal! what are you about there?"

Harold was immensely relieved at hearing his friend's voice, and started to him.

"I've got a prize. Come and help me to get it to the steamer."

Lewis came over to him at once, and Harold told him briefly how the seal had come into his possession.

"Indeed, this is a remarkable seal," said Lewis, on hearing the account. "I'll help you to get him to the steamer;" and picking the creature, which vainly tried to escape, up in his strong arms, he bore it off to the *Narwhal*, just as though it had been a tired child.

Captain Marling readily gave his consent to the seal being kept on board, and a big tub was at once filled with water and placed near the foremast, so that it could bathe whenever it wished. Many a time in the long and dreary days that were yet to come did Harold feel thankful for the good fortune which threw Shiney in his way, for he made a most interesting pet.

CHAPTER XIII.

GATHERING CLOUDS.

THE seal-hunt had been very successful, over five thousand pelts having been taken, and one of the largest tanks had been filled with the blubber.

"Now if I can only get a few walrus," said Captain Marling, reviewing with deep satisfaction the results of the day's operations, "I shall be very well content."

He had not to wait long before his wish in this direction was gratified also. They were in a very good place for walrus, the fish being exceedingly plentiful, and these ugly monsters were frequently sighted from the crow's nest. They proved very wary and difficult of approach, however, and for some time all attempts upon them were fruitless. At length the tide of fortune turned in favour of the hunters. One day the look-out reported a number of black objects, looking very much like walrus, as lying upon the ice-field a good distance from the water, and therefore easily to be attacked. Captain Marling ordered out three boats, in one of which he went himself, taking Harold with him; and filled with men eager for the fray, they pulled rapidly toward the spot indicated by the look-out.

Landing upon the ice in the most cautious way possible, the hunters, leaving two men in charge of each boat, spread out in a long line, and advanced upon their prey. The huge creatures were sunning themselves in a level portion of the field about two hundred yards from the edge of the ice. Fortunately the wind was blowing from them toward the men. Had it been otherwise the presence of the intruders would have been detected at once, even though they themselves had been invisible. Creeping along on all-fours, and using every inequality in the ice to conceal their approach, the hunters got within fifty yards of their game without being detected. Their concealment was no longer possible owing to the level character of the ice, and rising to their feet they cocked their guns, and drawing their lines close, hastened to the attack.

They were well prepared for the fray. The captain, Lewis, the surgeon, and the engineers, all carried Winchester repeaters of the best make. Even Harold had a small rifle of the same kind, which he had learned to handle very cleverly under Lewis's careful tuition. Several of the men carried rifles and shot-guns, and the rest had harpoons and lances, so that the chances of the prey did not seem very bright as this line of well-armed men moved down upon them. But they did not show the least alarm. On the contrary, they faced their foe with a look that seemed to say, "Who are you? and what do you mean by disturbing us in this rude manner?"

The appearance of men was evidently a novelty to them, and they would consider the situation a moment before deciding whether to fight or flee. They were fifteen in number—four huge bulls, as many young calves, and the rest cows. They were extraordinarily grotesque and gruesome-looking creatures, as they lifted their heads and stared at the oncoming men, sniffing in a fierce way that showed their uneasiness. They much resembled exaggerated seals in body, but their heads, with the long, shining tusks, the forest of bristles standing out all over the nose, the sharp, wicked eyes, had little of the seal about them, and Harold involuntarily shrank behind his father, say-ing, "What awful brutes! They'd kill the whole of us if they got a chance."

"Fire at the bulls; never mind the others," cried the captain, and the next instant the crack-crack-crack of the Winchesters told that the battle had begun. The bullets were well aimed. All four of the bulls were wounded, but not one of them fatally; and while the cows and calves, terrified at the report of the rifles, made off as fast as they could go, pursued by some of the sailors, the males, maddened by their wounds, charged down upon their foes with such vigour as to cause an immediate stampede. Huge and clumsy as they were, their immense strength enabled them to get over the ice with surprising speed; and could they have turned as rapidly as they could go ahead, they would have been awkward customers to handle.

As it was, the captain had a narrow escape from them, for in turning suddenly to avoid their charge, he tripped and fell headlong, his gun flying far out of his reach.

"Great heavens! the captain's down!" cried Lewis, the first to observe his chief's danger; for a great bull, looking terrible in his rage, was almost upon him. Quick as thought he raised his rifle. But a cartridge had jammed, and it missed fire. Swinging it by the muzzle, he sprang forward, intending to use it as a club, when the sharp crack of another rifle rang out, and the furious monster, stung in the neck by a well-aimed bullet, turned with a roar upon its new assailant, giving the captain time to scramble to his feet and regain his rifle.

Once it was in his hands, a succession of shots, fired as fast as the gun would work, was poured into the walrus's broad breast, and he fell helpless upon the ice, bleeding from a dozen wounds. Not until then had he time to see who it was whose timely shot had distracted the walrus's attention at that critical moment when its terrible tusks seemed to be almost right over him; and what was his surprise and delight to find that it was his own son, who, rifle in hand, came up, exclaiming, " Hurrah, father! I saved you. It was I shot him in the neck."

" God bless you, my boy!" cried the captain, throwing his arms around Harold and giving him a hug worthy of a bear. " You did, indeed. But for that shot the brute would have had me sure. But now we've got him, and you shall have the tusks for your good shot."

The tusks were very fine ones, fully two feet long, and Harold felt very proud of his prize, as well as of his timely assistance to his father. By this time two more of the bulls had been slain ; but the fourth, the youngest and most active of all, had made off toward the water with such celerity, although already wounded in more than one place, that ere its progress could be checked it had reached the spot where the boats lay in waiting. Then it caught sight of the boats, and saw its opportunity for revenge. Plunging into the water as if to disappear altogether, it rose suddenly at the stern of one of them, and lifting its head dealt a powerful blow with its tusks, crushing through the light timbers as though they had been paper, and causing the men, who were quite unarmed, to spring out on to the ice with amazing celerity.

But in seeking revenge it found its own destruction, for its tusks were so embedded in the tough wood-work of the boat that it could not withdraw them, and one of the engineers coming up at the call of the boatmen, put half-a-dozen bullets in its head, thus completing a fine day's work.

The hunt had been entirely successful. All four of the big bulls and the same number of the cows lay stretched upon the ice, the calves being allowed to get off scot-free. The bulls were enormous creatures, weighing not less than a ton apiece, and their blubber would be well worth the trouble of flinching them, while their tusks and hide would each be of great value in their way; so that Captain Marling was in great humour, as he well might be, at the results of their first walrus-hunt.

From Repulse Bay, where she had made so pleasant and profitable a stay, the *Narwhal* sailed out through Lyon Inlet, and past Baffin's Island into the broad waters of the Fox Channel, now fairly clear of ice. Here Captain Marling hoped to secure a couple more whales, and if they were good large ones he would be content, as with the whale blubber already secured, and that from seals and walrus subsequently added, his tanks were pretty full, and the financial success of his expedition placed beyond a doubt.

He had not yet published his plan already mentioned to his men, and thought that they knew nothing about it. He quite expected some demurring on their part; for although they had shipped with him on the understanding that the voyage might continue for eighteen months, should he so desire it, still, in view of the full tanks, and of the natural eagerness of the men to realize upon their shares, besides the equally natural reluctance to tempt the unknown perils of an Arctic winter, he felt sure they would do their best to persuade him to turn about after the Fox Channel had been hunted over, and drop down past Southampton Island into Hudson Strait and thence homeward.

Strict injunctions had been laid upon those whom he had taken into his confidence not to say one word until he deemed the time had come, and so far he had seen no reason to imagine that these injunctions had been disobeyed. Nevertheless they had been, and evil counsel was already at work among his men, sowing seeds of dissension that were to bear the black fruit of death in the near future.

But the captain knew nothing of this, and sailed straight on up Fox Channel, or took a slant to right or left, according as he thought the chances would be best for whales. The good fortune which had befallen him in Rowe's Welcome did not desert him; for three days after entering the channel a regular school of whales was sighted from the crow's nest, and every one of the boats was despatched in hot pursuit. Harold did not accompany Lewis this time, for although his father had not forbidden him to do so, he knew he would feel more comfortable if he remained on board, and as it turned out he had good reason to congratulate himself on his action.

The whales were at least a dozen in number, and when first sighted were more than a mile from the steamer.   But their course led them nearer, and by the time the boats reached them they were not more than half a mile away, so that Harold, perched up in his favourite eyrie in the main cross-trees, had a fine view of all that went on.   The six boats had started close together, and were now spread out in a sort of semicircle, advancing swiftly upon their prey, which were still ignorant of their approach.   This ignorance did not last long, however.   The two outside boats were nearest their fish, and presently two sharp reports rang out, the one right after the other, and the ever-welcome cry of "A fall! a fall!" came from each as the well-aimed harpoons buried themselves in their living targets.

Instantly the other whales dived out of sight, as did those to which the boats were fast, and there was nothing but the swishing water to show where, a moment before, the great cetaceans had been displaying their bulk.

"Give way, all," shouted the steersmen of the loose boats, sweeping their light craft around in pursuit of the vanished monsters.

"Stand by to watch the line!" cried the harpooners of the fast boats, as their lines ran out at smoking speed.

Excitement reigned supreme, but there was no lack of discipline, notwithstanding. Every man knew his work, and gave his whole attention to it. After an interval of trying suspense, the whales began to rise again in different places, and soon the cry, " A fall ! a fall !" announced that another boat had been successful.

" They have three whales already, father !" screamed Harold, in delight.

" Ay, ay, my boy," replied the captain. " If they only bring them all safe alongside, we won't need to do any more whaling."

They did eventually succeed in bringing them all alongside, but not until a struggle that came near having a lamentable termination for some of those engaged.

Three of the boats being fast to good big fish, the other three thought it best to come to their assistance, especially as the rest of the whales had made off at such a rate that it would have been useless to follow them. One of the fast boats was Collins's—the dark-browed, taciturn Newfoundlander—and he was evidently having such trouble with his prize, that Lewis, having failed to get a fish for himself, rowed up to his aid. Twice had the great creature " sounded," taking almost the whole length of line in his mad efforts to shake off his assailants, and each time had reappeared at such a distance that ere the boat could row up to lance him he had time to recover his breath and dive again. For the third time he rose, spouting out a high column of water, deeply tinged with blood; and now, happening to catch sight of the boat, he made straight for it, holding his vast mouth wide open, as though he would engulf it at one mighty swallow.

Seeing the extremely perilous position of Collins's boat, Lewis—who was lying at some little distance, so as to be ready to afford assistance if necessary—sent his boat flying toward the whale, and while it was yet twenty yards from its intended victim, hurled a hand-harpoon with splendid aim into a vital spot. The pain of this fresh attack caused the furious monster to swerve sufficiently from its course to miss the boat at which it was aiming; but just as it passed it threw its tail high up into the air, and brought it down with a terrible crash, striking the boat across the bow and crushing it in as though it had been an egg-shell. The men were all thrown into the foaming water, but, strange to say, not one was hurt—Collins having a remarkable escape, thanks to a lucky dodge out of the way as the destructive tail struck the boat. It had been better, both for him and his shipmates, if he had not escaped. They would, at least, both have been spared the dark trouble that was drawing near.

Lewis's boat was at hand, not only to pick up the men, but to continue the fight with the whale; and having accomplished the first, set to work at the second with such vigour that in fifteen minutes more the struggle was over, and the prize made ready to be towed back to the steamer. One by one the boats returned to the *Narwhal* ere darkness set in, and before all hands were piped to supper, three large, fine whales were moored safely alongside as trophies of the day's struggle. Captain Marling deemed this a timely occasion upon which to declare his prepared programme of future action to his men. Accordingly, after supper, when they might be expected to feel in thoroughly good humour—being well rested and fed—he summoned them all to the quarter-deck, and in the course of a brief address outlined to them his plans for the coming winter.

To his great surprise, the men showed unmistakably by their manner that this declaration was no surprise to them. They were, evidently, not only prepared to hear it, but prepared to oppose it too ; and the captain had hardly finished speaking before one of the steersmen—"Big Alec," as he was known among his shipmates, with whom he was a recognized leader—stepped forward a little, and on behalf of the crew presented, respectfully enough, their objections to the captain's proposition.   They were, in brief, as follows :

"When they shipped on board the *Narwhal*, they had understood that the principal object of the voyage was the whaling ; and complete success having crowned their efforts, they thought they ought to return and reap the benefits of their good fortune.   It was true that their articles bound them to serve for two years, if the captain desired, but they supposed that provision was to be enforced only in event of an unsuccessful season at whaling.   There was nothing for them to gain by a winter in the north, and they were anxious to return to their families that autumn."   Then, emboldened by the attention with which the captain listened to him, Big Alec went on to say so much about the terrors of an Arctic winter, that Captain Marling, who knew well enough that he was not speaking from personal experience, suddenly interrupted him with the question, "What do you know about it ?   Who told you all that stuff, sir ?"   Taken aback by the unexpectedness of the interrogatory, Big Alec betrayed the whole matter: "Why, Mr. Collins, sir, and Mr. Stacey says it's all true too."

Instantly the quick-tempered captain was aflame with indignation. "Aha!" he cried. "That's the kind of work that's been going on, eh? So you've been listening to those hang-dog rascals that I was fool enough to take on board at Newfoundland? Very well, sirs. I know how to deal with you and them. Go down below, and let me hear no more of your nonsense." And turning his back upon the men, who looked crestfallen but obstinate, Captain Marling retired to the saloon in high dudgeon. He was no less perplexed than indignant; for this interference of Collins and Stacey in his cherished plans might have very serious consequences, if not properly dealt with, and he was not quite clear in his mind as to the best course to pursue. All this made him feel unusually depressed, and his gloom communicated itself to the others, who soon learned what was the matter; so that the usually cheery gathering in the saloon was under a cloud that evening, and the morrow was awaited with much anxiety by all.

CHAPTER XIV.

THE STORM BREAKS.

CAPTAIN MARLING spent an anxious, troubled night; but by morning he had quite decided upon his course of action, and proceeded to carry it out with his habitual promptitude. Immediately after breakfast he called the crew before him in the waist of the ship, and spoke to them as follows:—

"Shipmates, I am more surprised than I can tell you that you should try to play me false in this way. Have I not always done the square thing toward you—paid you good wages, given you good grub, treated you like men? And now I find that you have been opening your ears to the talk of those false traitors, Collins and Stacey, and you want to turn back and go home instead of following your captain as you have always done before. I am astonished, shipmates, that you should want to treat me in such a rascally way; and now that I know who is at the bottom of the mischief, I am going to serve them out as they deserve. Before you all I order those two men to be placed in irons for exciting insubordination on board ship, and to be kept in confinement until I shall see fit to release them. That's my way of dealing with traitors. As for you, men, go to your work, and let me hear no more from you."

There was a murmur of surprise and protest from the men, and had they been given the chance they would no doubt have spoken; but directing Strum and Lewis to carry out his orders with regard to the Newfoundlanders, Captain Marling turned away and went back to the quarter-deck, where he took up his position near the stern, looking moodily out over the water. It had been better perhaps if he had given the men an opportunity to reply, and undertaken to reason with them a little. But he felt so hurt and indignant that they should have listened for one moment to their evil advisers, that he had no patience to argue with them, and so determined to exercise his authority as captain to the fullest extent.

Collins and Stacey offered no resistance to being placed in confinement, although the former was heard to mutter under his breath that that "coxcomb of a captain would suffer dearly for this." They knew well enough the time was not ripe for desperate measures, and that there was no alternative save to submit, which they did with lowered brows and darkly-gleaming eyes. They were confined amidships in a sort of store-room which had recently been emptied—not by any means an inviting place, but good enough, the captain thought, for such characters.

Their advisers having been removed, the men went back to their work, but in a sullen silence that boded no good. They did not seem at all like themselves. They had three splendid whales to flinch and stow away, and ordinarily this occupation would have been attended with song and joke and hearty laughter as the work went merrily on. But now they had neither song nor joke nor laughter. A heavy pall hung over them, and, as a consequence, they did not work one half so fast or well. Captain Marling saw all this clearly enough, but made light of it.

"They've just got a fit of the sulks," said he to Lewis. "They'll soon get over it if we leave them alone."

Busy as all were, the day seemed to pass with extraordinary slowness. Harold, who had only a dim idea of what the trouble was—for his father concealed its magnitude from him so far as possible—

found it terribly dull on board. Everybody seemed preoccupied and disinclined to be sociable, and but for the companionship of Patsy and his pet seal, Shiney, who was already beginning to be very much at home and exceedingly entertaining, he would have been sorely at a loss to know what to do with himself. Shiney proved a great comfort. He showed quite an appreciation of fun, and evidently enjoyed a romp, splashing in and out of his tub at a great rate, and playing many cute little pranks.

Evening brought with it no easing of the situation, and it was clear that the captain and his associates in the saloon felt the state of affairs to be very serious as they gathered together for consultation after tea. To face the privations and perils of a winter in the ice with a mutinous crew was certainly not an attractive prospect, and yet to be balked of a long-cherished project, when just on the threshold of its accomplishment to be compelled to turn about and give it up—this was something against which one's manhood sturdily rebelled. It was a long and earnest consultation that took place, and Harold, profound as was his interest, fell asleep before it was finished— the decision ultimately arrived at being to maintain a firm front against the men, dealing as sharply as might be necessary with any who showed further signs of insubordination.

The next day, which was spent in "making off," passed pretty much as had its predecessors, the work being done in silence, with a general air of uneasiness prevailing throughout the ship. Harold and Patsy, who had both by this time got a pretty clear inkling of the trouble, talked much together about it, and more than once crept cautiously down to the dark store-room, where the Newfoundlanders were imprisoned, to take a peep at the men who had behaved so badly and betrayed the captain's confidence. Once when they went they found Big Alec engaged in earnest conversation with the prisoners, but although he was evidently very much flustered at their seeing him, and slunk off immediately, they did not think the incident of any importance, and said nothing about it. Had they mentioned it to the captain, the tragedy that was near at hand might possibly have been averted.

After enjoying freedom of action for more than a week, the *Narwhal* became beset by an ice-pack which surrounded her in all directions, and she was evidently a prisoner until there should be a decided change in the wind. The evening being damp and chilly, no one felt tempted to linger upon deck, and after darkness had settled down upon the steamer all was quiet on board, there being nobody visible save two seamen doing duty as the bow watch, crouched in the shelter of the foremast, and Peter Strum leaning silently over the stern taffrail. In the saloon the captain, the second mate, the surgeon, and one of the engineers were seeking a diversion of their thoughts in a friendly game, and were growing deeply interested. The other engineer was making additional entries in the journal he delighted to keep, and Harold was reading for the third time one of his favourite books, to wit, "Mr. Midshipman Easy." It seemed a very peaceful and secure scene, and little would one have thought how soon it was to be rudely disturbed.

It was Harold's custom since Shiney had come into his possession to run up and say "Good-night" to his pet before going to bed. Suspecting nothing, he went up as usual this night, and finding the seal in his accustomed corner near the foremast, was bending over it caressing its soft, sleek head, when suddenly he was caught in a powerful grasp, a huge horny hand was thrust over his mouth so that he could not cry out, and before he realized what was

being done with him, he was borne off by Big Alec
to the forecastle, and dumped down in the midst of
the sailors with the stern injunction : "Keep your
mouth shut now, or something'll happen to you."

Startled and bewildered, Harold looked about him,
and among the first his eyes fell upon were Collins
and Stacey, whom he supposed to be lying in irons in
the dark store-room. Clearly there had been treachery
somewhere, and he was at once reminded of seeing
Big Alec sneaking away from the place of the Newfoundlanders' confinement. He further noticed that
there was a good deal of drinking going on among
the men. They must have got at the spirits somehow, for the brandy flowed freely, and all seemed
to be partaking of it. Collins, who appeared to
be the leader, showed great satisfaction at Harold's
capture.

"Aha !" he said, with a wicked laugh. "Now
we've got the young cub, perhaps the old bear won't
growl so loud."

Captain Marling did not notice Harold going out,
but after he had been gone some little time, he looked
up from his game, and missing him, exclaimed,
"Hello ! what's become of Hal ?"

"Gone up on deck to say good-night to Shiney, no

doubt," answered Lewis. "He always does it before he goes to bed."

"Well, it's about time he was in bed," said the captain. "Hi, there! Patsy. Run up on deck and tell Master Harold that I want him."

Patsy hurried off to obey, and the game was resumed. Ten, fifteen, twenty minutes passed, and Patsy did not return.

"Plague the brat, what's keeping him so long?" said Captain Marling testily, and ending the game he was just about to go up on deck himself, when Patsy rushed into the saloon, his face as white as a sheet, and his whole appearance betokening great agitation.

"Hello, youngster, what's the matter?" cried the captain, in surprise.

"Oh, sir, there's a great deal the matter," repeated Patsy. "They've got Harold in the forecastle, and Mr. Collins and Mr. Stacey are there too."

"Harold in the forecastle, and Collins and Stacey free! Great heavens! what's the meaning of this?" exclaimed the captain, a sharp thrill of apprehension striking to his heart. Then he added, more calmly, "This is bad business, shipmates; we must look into it at once. Let us get our revolvers."

There was some slight confusion as each one

hastened to his cabin and returned revolver in hand. The faces of all looked very grave. A crisis whose magnitude it was impossible yet to estimate was at hand. The future hung upon what might happen within the next few minutes.

"Stay here a moment while I go on deck and see how the land lies," said Captain Marling, as he quietly left the saloon.

After a brief absence he returned, his face looking graver than before.

"Things look pretty bad," he reported. "The men have got hold of the spirits somehow, and many of them are in liquor already. God knows what mischief they'll be up to. I wish to heaven my boy were here. But surely they won't do him any harm."

They were not left long in doubt as to what the men had in mind, for presently there was a tramp of heavy and unsteady feet along the deck and down the companion-way, which soon explained itself by the entry into the saloon of fully a score of the sailors, with Collins and Stacey at their head, while Big Alec, holding Harold fast in his mighty grasp, was not far behind. There could be no mistake about the purpose of the intruders. They had come to bend the captain to their will, and in view of the penalties to which they exposed themselves by thus breaking into open mutiny, had fortified their courage by potations of brandy until they felt equal to any excess.

There was a look of dark, leering triumph upon Collins's evil face as he surveyed the captain's little band that, numbering only six in all, now stood together in the rear of the saloon, and then glanced back over his own supporters, of whom there were more than three to one, while yet others were on deck awaiting the issue. His imprisonment had filled him with a wild passion for revenge, and he was determined to have it, regardless of all consequences.

" Well, captain," said he, first breaking the ominous silence, " we've just come down to say that we don't hanker after going any farther north, and that we think it's about time to right about ship, and make for home.—Isn't that about the size of it, shipmates? "

A murmur of assent from those behind expressed approval of his words, and with an insolent smile he awaited a reply. It came without delay.

" I am the captain of this ship," said the captain, calmly yet sternly, " and will suffer no one to dispute my authority. By this action you have become guilty of mutiny; and I give you warning, all of you, that unless you leave this cabin at once and go back to your berths, I will hand you over to the authorities at the first port we reach."

Collins was evidently ready for this, and it had not the slightest effect upon him; although some of the more sober of the men winced slightly, and began to look as if they wished they hadn't come on any such errand.

"Thank you, captain," answered Collins, still smiling insolently. "I haven't a doubt you'll do exactly as you say, if you only get the chance. But we'll take mighty good care you don't. And what we've got to say is just this: If you don't consent to turn about and go home, and promise to say nothing about this, we're just going to give you and your friends one of the boats, and let you go on up north as far as you please, while we'll take the steamer, and go down south as far as we please. That's our little plan. What do you think of it?"

A strange chill went to Captain Marling's heart as he heard these words, and there came up in his memory the talk he had had with Harold about Hudson and his cruel fate. But there was not a sign of fear or irresolution on his countenance as he made reply : "You black-hearted scoundrel! This is all your devilish doing, and dearly shall you pay for it. Leave the cabin this moment, or I'll shoot you where you stand." And the shining barrel of a revolver was pointed straight at Collins's head.

Taken completely by surprise, the rascal fell back a pace or .two, his face livid with rage and fright. As he did so, Big Alec stepped forward, holding Harold by his left hand, and in his right one of the long, keen knives used in separating the whale blubber from the skin. "Shoot, if you dare!" he shouted—for he was wild with drink—"and your cub shall pay for it."

Poor Harold gave a shriek of terror, and cowered down as the half-crazed giant held the wicked blade in his face. Quick as a flash all six revolvers of the captain's little band were levelled at the mass of men in front. A moment more and there would have been an awful scene; when suddenly—from no one knew where, for no one saw his approach—Patsy sprang forward, his face aflame with heroic purpose. In his hand he held a short iron bar which he had picked up somewhere, and ere Big Alec could avoid him, indeed before he was aware of his proximity, he brought the bar down with all his might upon the arm that held Harold, shouting, as he did so, "Run, Harold, run, for God's sake!"

With the spring of a deer Harold obeyed, and with the roar of a wild beast Collins rushed forward to intercept him. But ere he could take a second step a sharp crack split the air, and with a bullet in his brain the ill-starred wretch pitched forward at the captain's feet—stone-dead.

The report of the revolver was followed by an instant of profound silence. Appalled at the fate of their leader, the mutineers stood as if petrified; and seizing the opportunity, Captain Marling, still keeping his revolver pointed at them, while they shrank back in terror from its deadly muzzle, shouted in his sternest tones, "Fools! Do you want to share the fate of this scoundrel? Back to your berths, now. I'll talk to you in the morning, when you've got your wits again."

Completely sobered by fright, and abject in their collapse, the men one by one slunk out of the saloon until only Big Alec was left. He was about to follow, when by a sudden impulse he wheeled round, and with all his drunken fury turned to maudlin penitence, rushed across the saloon, and throwing himself on his knees before the captain, while the tears poured down his bronzed cheeks, cried out imploringly, "For God's sake, captain, forgive me! I didn't know what I was doing."

Captain Marling, who in his delight at Harold's safe deliverance, and relief at the happy turn of affairs, felt as if he could forgive everybody and everything, said very gently, "All right, Alec; we'll see about it in the morning. Just call one of the men, and put that thing out of sight," pointing to Collins's body.

Big Alec went off, and bringing back another sailor, the two carried away the body, and then departed again without another word.

They were very quiet in the saloon when once more left to themselves. The gravity of the peril, and the tragedy by which it had been averted, made them little inclined to talk, although their hearts were filled with profoundest thankfulness.

"My darling boy!" said the captain at length, patting his son's curly head, "little did I dream into what dangers I was bringing you when I consented to let you come. It must be Aunt Etter's prayers that keep you safe from harm. God bless the dear woman! What will she say when she hears of this?

But, bless my heart, I've never thanked Patsy. Where is the boy?—Hi, there! Patsy! Patsy!" he shouted at the top of his voice. Divining the reason of this stentorian summons, Patsy emerged from the pantry, blushing furiously. "Come here, my boy, come here, and get a father's thanks," cried the captain. And then, as the little stowaway came and stood before him, he took his two hands in his, and pressing them warmly, said: "Now, shipmates, bear me witness. So sure as my name is John Marling, Patsy Kehoe shall never want for anything while I'm alive; and if we all get safe back to Halifax again, I'll take good care that he shall want for nothing after I'm dead either.—God bless you, little man! it was a lucky day when you stowed away on board the *Narwhal.*"

CHAPTER XV.

DUE NORTH.

A VISITOR on board the *Narwhal* the morning after the tragedy would have had to be very unobservant if he did not soon find reason for thinking that there was something amiss. From Captain Marling downward, everybody seemed uneasy and restless. The sailors went about their work in silence, and wearing a dejected, downcast air that would have been sufficient of itself to arouse inquiry. The officers talked much together, but in a quiet, earnest way that showed the subject of their conversation to be of grave moment; and the captain spoke to no one save when he chanced to pass near Harold, and then he never failed to lay his hand affectionately upon the boy and make some pleasant remark to him. Even Harold had been so deeply impressed by the events of the preceding night, that a shadow had fallen upon his young spirit for the time, and his merry whistle was unheard. He had never before been in the presence of death, and the awfully sudden fate which Collins had brought upon himself was a great shock to him. Indeed, through all his life the memory of that scene in the saloon would never lose its vividness. His own peril, the flashing knife in the hand of his drink-maddened captor, Patsy's brave blow, and his own wild dash for liberty, Collins's furious spring after him, and then the sharp crack of the revolver and the dull thud of the traitor's body upon the floor—it took but a moment, but oh, how dreadful it all was!

Early in the forenoon Captain Marling called the crew together, and committed the body of the unhappy Collins to the icy water, wrapped in a canvas shroud and heavily weighted, that it might sink far down into the silent depths. He could not suppress a sigh of relief as the white, shapeless thing slipped into the water and vanished. It seemed as though it carried with it all the trouble that had been weighing so heavily upon him of late; for it not only removed the chief cause of that trouble, but made the way more open for him to deal leniently with those whom that evil man had led astray.

For a brief space after the corpse with a soft splash

had disappeared for ever there was a dead silence on board, during which each man might almost have heard his neighbour's heart beat. Then, lifting his head and looking into the faces of his men with an expression full of mingled kindness and reproach, the captain asked them, "Shipmates, have I not always been a good captain to you?"

"Ay, ay, sir!" came in hearty cheers from the men.

"Did I ever refuse to listen to any of your reasonable requests?"

"No, sir! no, sir!" cried the men in chorus.

"Well, don't you think you have treated me very badly in listening to that poor wretch who has paid so dearly for putting evil thoughts into your heads?"

"We have indeed, sir, and right sorry we are for it," spoke up Big Alec, in tones whose sincerity was unmistakable, while the others murmured their assent.

"I believe you, my men, I believe you," continued the captain; "and because I do, although you have been guilty of the worst crime a sailor can commit, I do not intend to punish you as the law gives me power to do. Indeed, on one condition, I will not punish you at all."

"Please, sir, what's the condition?   Just name it, sir," said Big Alec eagerly.

"Well, it's just this," answered Captain Marling. "If you will all promise me faithfully to stand by me and the ship, no matter where I see fit to take her, and will go back to your duty with good hearts, ready for anything, I will say nothing more about what happened last night."

"We will, sir, we will, with all our hearts!—Won't we, mates?" shouted Alec, turning round to his fellow-sailors; and when they chorused their assent with equal heartiness, he took off his cap, swung it in the air, and called for three cheers for "Captain Marling! God bless him! and we'll follow him clear to the North Pole!" to which the men responded with a vigour that awoke the echoes amid the Arctic silences.

"All right, my men; it is a bargain between us. Bygones shall be bygones, and we'll be good friends to the end of the chapter," said the captain, his face showing plainly that the burden had been removed from his mind, and that his way seemed clear before him.   He dismissed the men, and looking very much relieved they went back to their work.   The crisis was over; the difficulty happily overcome.   The captain had won the day, and the only sufferer was the poor wretch who, forgetful of all the captain's kindness, and regardless of his own pledged duty, had tried so hard to play him false.

There was a change in the wind about mid-day, and the ice-pack began to disintegrate, allowing the *Narwhal* to make some progress northward. But it was very slow work at first, and required careful management. The " leads," as the open water between the floes are called, were few and far between, and often after following one to its end, it would be found to be stopped by an iceberg or a specially heavy mass of floe ice which put further advance in that direction out of the question, and rendered a retreat necessary.

Captain Marling grew increasingly impatient as that day passed, and another and another, without the stubborn ice showing any disposition to abate its opposition. He was anxious to get as far north as possible before the brief Arctic autumn came on, as the navigation of Fury and Hecla Strait would in all probability be very difficult, and it would not be wise to attempt it in a stormy season. By a free use of steam he might have got ahead a little farther, but this was just what he wished to avoid. He had sufficient coal left in his bunkers to carry him to the end of his expedition provided it was sparingly used; but if he squandered it now in fighting the floes, he would have to do without it when fighting the cold at their winter quarters.

The delay, however, had one advantage: it gave him plenty of time to talk with Harold, and he told the attentive boy many an interesting story of the early explorers and their wonderful courage. They had such miserable little vessels in comparison with the powerful, well-equipped *Narwhal*, for instance, that their achievements were rendered a'l the more heroic. Harold was filled with admiration for Frobisher and Baffin and Master John Davis, those fearless souls who fought their way far into this wilderness of sea and ice long before the days of Franklin and M'Clintock and Parry.

"Why, do you know, Harold, they had to resort to the strangest plans to save themselves from destruction," continued the captain. "During one of brave old Frobisher's voyages a barque named *Dennis*, of only a hundred tons, mind you, was struck so hard by a huge ice-floe that she sank at once in the sight of the whole fleet, and soon after a tremendous tempest sprang up that played the very mischief with the other vessels. The ice surrounded them on all sides, and in their desperation, fearing every moment lest the sides of their ships should be crushed in, they protected them with cables, mattresses, planks, and even spare masts, which might act as fenders against the pitiless blows of the ice. A still stranger device was that employed by the *Judith*, another of the same fleet. When being sorely beset, she made fast to the biggest piece of ice she could find, and crowding on all sail, used it as a sort of battering-ram with which to force her way through the ice, and it seems to have been quite a success too."

"But, father," interrupted Harold, "nothing like that could ever happen to us in this great big steamer, could it?"

"Not likely, Hal, not likely," answered the captain; "but we've got to be very careful, all the same. There's always the danger of being nipped, you know, and no vessel that was ever built, not even the *Great Eastern* herself, could stand being nipped."

"Why, what's being nipped, father?" cried Harold, opening his eyes wide.

"Simple enough, my boy, simple enough. It's just getting in between two great ice-floes which the wind or current is pressing together, and then, unless something interferes, the ship is cracked just as a nut is

cracked in the nut-cracker.   It is a very serious business, I can tell you, and I devoutly hope we won't
have any of it.   But I must run up on deck now,
and see how we are getting along."   And leaving
Harold to think over the perils of northern navigation, Captain Marling went up on deck.

He found the prospect looking much more promising.   The ice was separating floe from floe, wide leads
were opening up in all directions, especially to the
·northward, and the wind was favourable, blowing
strongly from the west.   With his good spirits
thoroughly restored, the captain sang out his orders
cheerily, and under a full press of sail the *Narwhal*,
dodging this way and that way from lead to lead,
made good headway up Fox Channel toward Fury
and Hecla Strait.   With two days of ordinarily good
weather the entrance of the strait ought to be reached.

Although the tanks were now sufficiently well filled
to satisfy the captain, a sharp outlook was nevertheless kept for any game that might be met, as he
wished to make as complete a collection as possible
of Arctic trophies.   He was particularly anxious to
secure a good specimen of that strange fish the unicorn, or "unic" as the sailors called him for short;
and when one morning a fine "unic" was sighted

cutting through the water not far from the ship, he ordered a boat lowered immediately, and took command of it himself, with Lewis as harpooner.

Harold of course accompanied him. Wherever his father went he followed. There was no gainsaying that. The oarsmen were the best on board; and directing another boat to be made ready in case assistance should be required, the captain ordered his men to "give way," and off shot the boat after its prey. They had to approach the unicorn—or narwhal, which is its proper name—very cautiously, for they are wary creatures and easily startled, so that one must never forget the sailors' maxim to "keep off her eye." Pulling along smoothly and swiftly, the boat crept up to within striking distance, and then Lewis, not losing a moment, rose in his place, and using the hand-harpoon this time, sent it deep into the narwhal just behind the fin. With a tremendous start and flurry the stricken fish flung up its tail and dived into the depths, taking out more than a hundred fathoms of line as rapidly as any whale could do. Then it stopped all at once. The line lay loose in the water. The prize seemed already as good as won. Hauling the line in hand-over-hand, the men found little resistance offered on the part of their captive, and more than one-half of the rope had been coiled away in its place when the strain ceased altogether.   Noticing this, Lewis at once called out, " Look sharp there! the unic's coming up again!"

He had hardly uttered the words when, like an apparition, a huge form rose from the depths on the starboard side of the boat, there was a lightning-like rush that made the water fairly hiss, and then a crash as the mighty horn pierced through the tough wood as though it had been paper.   So fierce was the narwhal's charge that fully two-thirds of its horn entered the boat, the sharp point finding its way into the thigh of an oarsman who was sitting near the gunwale, knocking him off his seat, and inflicting a severe wound, from which the blood gushed out.

For a moment all was confusion.   Thinking only of the injured man, Captain Marling at once sprang to his side, and with that presence of mind which never failed him, whipped out his big silk handkerchief and tied it tightly around the poor fellow's leg, above the wound, thereby stopping the flow of blood, which otherwise would have been fatal.   In the meantime Lewis gave his attention to their plucky assailant, plunging a lance again and again into its broad back, for so firmly was its horn embedded in the boat's side that it could not extricate it. The keen lance soon did its work, and with its life-blood dyeing the sea around, the unicorn ceased to struggle.

"Hurrah!" cried Harold; "he's done at last."

"Yes," replied Lewis; "but he came pretty near doing for us first. By Jove, how he did charge us! I fairly shivered when his horn came through the boat." Then turning to the injured seaman, he said, "Much hurt, Bell? He hit you pretty hard."

"That he did, sir," answered Bell, who was holding on to his leg with both hands above the wound as the captain had bidden him. "A little more, sir, and he'd a' taken my leg off altogether."

"We must get back to the ship at once," said the captain, "and let the doctor look after this man's leg. Just signal for that other boat to come along."

Lewis signalled accordingly, and on the other boat coming up, the captain, the wounded sailor, and Harold got into it, and hastened back to the ship, leaving the boat with the narwhal in tow to follow more slowly. The horn had made a hole little larger than itself, and the boat did not leak badly, so that there was no danger of its filling. Arriving at the steamer, the sailor was handed over to Dr. Linton, who had him well bandaged up before the other boat got back, and painful as the wound was, no serious consequences were to be feared.

The question that now presented itself was how to get the narwhal on board without injuring his horn, which the captain was particularly anxious to secure uninjured, since the fish had made such brave use of it. After many plans had been suggested, none of which quite met the difficulties of the case, Captain Marling at length decided that he would rather make a bigger hole in the boat than have a break in the unic's horn. Accordingly the ship's carpenter was bidden to cut out the piece around the horn, and then the fore spek tackle being attached, the narwhal was hoisted on board in triumph.

He proved to be a very fine specimen, being almost twenty feet in length, and boasting a horn fully seven feet long, a really formidable weapon, which would have made it more than a match for the largest whale in the duels that these sea-monsters sometimes fight. The long horn was on the left side of the broad, blunt nose, and on the right side, hidden in the blubber that overlaid the skull, was a tiny horn scarcely a foot long, which went to show that this strange creature was not really a unicorn, after all. Harold put in a claim for this little horn, which his father very willingly allowed, and it made a valuable addition to the collection of trophies the young fellow was forming on his own account.

The two days of fine open weather Captain Marling prayed for had been granted him, and the *Narwhal* was now at the entrance of the strait which led into the mysterious region where he so daringly proposed to spend the winter. Under the best of circumstances the navigation of this difficult strait could not be otherwise than arduous. But if the fine weather held, the attempt was well worth making; so putting on all steam, and spreading plenty of canvas, the sturdy vessel entered upon her trying task.

Unfortunately the fine weather did not hold. The last of the two days which had been so helpful was what is called a "weather-breeder." It was a little too fine, and betrayed the near presence of foul weather, which indeed came the following morning, in the guise of a strong wind from the south-east, and dark, lowering clouds that foreboded the storm.

Those in the saloon were just sitting down to breakfast, when old Strum came stumping down the companion-way, looking more glum than usual.

"Well, Peter, what's the matter?" asked the captain, looking up and noticing the doleful expression on the first mate's face.

"Matter enough, sir: the strait's full of heavy ice," replied Strum.

"What!" exclaimed Captain Marling, springing to his feet; "you don't mean to say so? That's a bad look-out." And leaving his breakfast unfinished, he rushed up on deck.

A single glance was sufficient to confirm the mate's statement. Stretching from one cliff-bound shore to the other of the strait, and as far northward as eye could reach, was the dreaded floe ice pressing slowly southward upon the bosom of the current flowing resistlessly out of the Gulf of Boothia. Here and there its much-hummocked surface was broken by "leads," but they led nowhere, being always closed at the farther end. The prospect could not have been more unpromising, and with the wind blowing from the south there was actual danger. It was what sailors call a "strong ale wind," because it packs the loose ice tightly against the land floe, and ships are more liable then than at any other time to get nipped between the mighty masses and destroyed beyond remedy. When this happens, and the ship must be abandoned, the men rush aft and broach the casks of ale and spirit, thereby often forfeiting their own lives in their mad passion for strong drink.

Just such a wind was blowing now, and it blew all that day, rendering the utmost care necessary to save the steamer from being crushed between colliding floes. To make matters worse, the lowering clouds toward evening let fall a heavy snowstorm, through which it was impossible to see more than her own length from the steamer. Then were ominous and significant preparations begun to be made. Provisions were hoisted up from below, and ranged along the upper deck, so that they might be thrown into the boats or upon the ice at a moment's notice. The men were bidden to put on their warmest clothing, and to make up into little bundles that might be strapped upon their backs the things they absolutely required. The cooks were ordered to prepare a supper of the most substantial kind, and of this every one partook heartily, for it might perhaps be the last meal they would eat on board the steamer. Darkness set in early, greatly adding to the difficulties of the situation. Nobody thought of turning in. All held themselves in readiness to jump for their lives; for when a vessel is nipped, the relentless ice must pass either under or over the ship, unless it passes through her, crushing in both sides at once, as it has been known to do.

Beset by darkness, storm, and a foe against which nothing could be done, the *Narwhal* battled bravely with all three while the long hours of that anxious night dragged slowly on, and none could venture to prophesy what the next minute might bring forth.

CHAPTER XVI.

INTO WINTER QUARTERS.

THAT was an awful and ever-memorable night. The wind shrieked madly through the rigging, as though the spirits of the strait, resenting the intrusion of the *Narwhal*, were crying out for her destruction; the snow beat fiercely on the faces of Captain Marling and Peter Strum, as standing upon the bridge they peered anxiously into the darkness, vainly striving to gain some idea of what was before them; at intervals only too frequent the great cruel floes would crash together, and the stout frame of the steamer would moan and groan in their terrible embrace. Fortunately they seemed much broken up, probably in the passage through the narrow portion of the strait above; and although big enough to be fatal to a smaller and frailer vessel, the *Narwhal*, built of the best oak and teak, and specially strengthened with iron, came off unscathed from attack after attack, although no one on board could tell but that at the next nip she would share the fate of the *North Britain*, whose surgeon, sitting in the cabin, beheld the ice breaking through from both sides at once, and had barely time to beat a retreat; or of the *Laurel* and *Hope*, that were squeezed perfectly flat, and then thrown upon the ice, to sink helplessly to the bottom so soon as the pack loosened.

Fully alive to the perils of their position, Harold kept in the saloon, it being altogether too stormy for him on deck—his little bundle, containing a change of clothes and a few other necessaries, lying upon the table, where it could be grasped in a moment. Patsy kept him company. Ever since his gallant action at the time of the mutiny, the two boys had been greater friends than ever, and Captain Marling, who was nothing if he was not grateful, encouraged the lad to spend his leisure time in a corner of the saloon, lending him books to read, and in other ways showing a warm and kindly interest in him.

There had been so wonderful an improvement in Patsy since his uninvited appearance on board the steamer, that it would be doubtful if any one of his old companions in the Halifax slums would recognize him. The blessings of good food, proper clothing, and a comfortable bed had not been wasted upon him. He had grown both taller and stouter, and really good-looking. Instead of slouching along in a hang-dog way, he walked with as alert and springy a step as Harold himself. So quick was he to learn his duties, so faithful in discharging them, and so respectful at all times, that, recognizing the value of his services, the captain had some time back added his name to the ship's company and allowed him good wages; which fact, more than anything else, had caused the happy boy to respect himself, and to feel that he was at last of some good in the world.

The boys were sitting together in the saloon, doing their best to keep up one another's courage by taking the most cheerful possible view of the situation.

"If the worst does come, Patsy," Harold was saying, "and we have to leave the ship, we'll still have a good chance of getting off all right. Father was telling me a little while ago that in one year there were twenty whalers crushed to pieces up in Melville Bay, and nearly every man got safe home some way or other."

"But sure, Master Harold, we're not going to pieces at all," replied Patsy stoutly. "Didn't I have a dream last night that I was walking the streets of

Halifax, feeling as proud as the captain himself—God bless him!—with a fine suit of clothes on my back, and my wages clinking ag'in' one another in my pockets."

Harold, anxious as he felt, could not keep from smiling at Patsy's earnestness.

"And how do you know that your dream must come true ?" he asked.

"Because," answered Patsy, his voice sinking into a whisper, "I put my beads under my pillow, and sure the dream must come true."

Harold had too much regard for Patsy to smile this time. Indeed, he felt no inclination to do so; on the contrary, he took comfort from the simple faith of the stowaway, and it was with entire sincerity that he said, "I'm glad you told me about your dream, Patsy. I believe it's going to come true myself."

Patsy's eyes glistened. He had felt timid about giving his reason, fearing he would be laughed at, and Harold's respect for his way of thinking touched his heart, encouraging him to add, "And sure, Master Harold, as soon as I do set foot in Halifax I'm going to the church, and I'm going to give particular thanks for our getting out o' this scrape."

The boys grew weary as the night wore on, and Dr. Linton having promised to waken them the moment there was immediate danger, they curled up one at either end of the long sofa, and soon fell asleep, to dream, perhaps, that they were both in Halifax again, with all the perils of the voyage past.

When they awoke it was broad daylight, and the sturdy *Narwhal* was still afloat and uninjured. They went up on deck together, and found the captain and Strum at their post, but not in at all as anxious a state of mind as they had been during the night. The snow-storm had ceased. The sky gave token of fine weather being not far off, and the ice was much less threatening in appearance.

"Hello, Hal! up already?" was his father's cheery greeting. "The old ship has weathered it all right, you see; but we've had a pretty hard time of it up here, I can tell you."

"Yes, indeed, father. I felt so sorry for you out in the storm all night. I hope when I come to be a captain I won't have to pass many nights like last night," said Harold.

"You've got to take whatever comes, my boy, fair weather or foul, whether on sea or land, if you ever expect to accomplish anything," answered the captain.—

"I think we've got through the worst of it now; and if you'll just take charge of the ship," he continued, addressing Lewis, who was standing near by, "Peter and I will go below and take a little rest."

They had got through the worst of it. Every hour the prospect improved, and by mid-day they were able to push on through the strait at a rate of speed that would bring them out into the Gulf of Boothia ere nightfall.

The scenery through which the *Narwhal* passed was very grand, albeit somewhat monotonous. Great dark cliffs rose rugged and frowning from the shore, broken here and there by coves and fiords, in which the remnants of last winter's ice still lingered, being already reinforced by fresh additions, for the nights were growing cold now. Where the cliffs opened a little, allowing a peep into the interior, faint patches of green yet marked the hillsides, and more than once the telescope revealed the presence of reindeer feeding upon the mosses and Arctic grasses which form their scanty food.

All that day the steamer forged ahead with but little interruption, the ice being so much broken up that she could easily force her way through it, the wind meanwhile blowing gently from the west. The biggest whale that ever disported its vast bulk in Arctic waters would not have tempted Captain Marling to turn aside in its pursuit. The end of October was at hand. The steamer must be securely settled in her winter berth ere November came, for it could not very well be done after that; so crowding on both sail and steam, the captain made the *Narwhal* put her best foot forward, so to speak, and the good ship covered herself with credit.

The sun had yet a little way to go before disappearing for the night, when amid the cheers of the sailors the steamer passed out of the strait into the broad waters of Boothia, and the most difficult part of her work was finished. It only now remained to seek a safe harbour, and make ready for the long, cold winter that would soon be upon them.

Captain Marling had determined to try Garry Bay, which cuts into the Melville Peninsula near its head, and just around the corner, as it were, from the outlet of the strait. Accordingly, the *Narwhal* lay to that night, and at daybreak next morning steered toward the bay, which was reached the same afternoon. A careful examination satisfied him that he could hardly have made a better choice. The bay was deep and free from obstruction. To the north and east the cliffs rose up in solemn majesty, offering complete protection from the winds that were most to be dreaded; to the south the land lay low and level, while westward the prospect was uninterrupted as far as eye could see. Here and there the mighty line of cliffs was broken by valleys running far inland, through which the captain promised himself exploring forays that would serve to vary the monotony of winter life.

"Capital, sir, capital!" said he, rubbing his hands joyfully, and beaming upon the imperturbable Strum, as he took in the many advantages of the situation. "Here we'll be as 'snug as a bug in a rug.' No northerly or easterly winds to bother us. No current to disturb the ice. Good hunting-ground to the south, and plenty of 'Huskies' not far off, I'll be bound."

Strum grunted his assent. If the captain was satisfied he was too, and that ended the matter.

Pushing well up to the head of the bay, the *Narwhal* was brought to anchor in a sort of a natural dock formed by two projecting ridges of rock ere the darkness settled down.

The following day was devoted to a thorough cleaning and overhauling of the ship. From the stokers and oilmen down in the dark engine-room up to the sailors taking off the top-masts, every man worked with a will. The weather was favourable, but who could prophesy how long it would continue so? and the best advantage must be taken of it, for there was much to be done ere the vessel would be ready for the winter. The ship having been made clean and put in apple-pie order, the next business was to turn her into a house. This operation Harold found wonderfully interesting. Having had it in view from the initiation of his enterprise, Captain Marling had made due provision, and in the forehold there was a pile of planks, rafters, and joists, that now for the first time revealed their purpose. The deck was turned into a carpenter's shop; sawing, planing, and fitting piece to piece went on. All who could be of any use lent an assisting hand, and the work progressed rapidly.

Always quick to think of anything that would impart a little variety to the life of his sailors, Captain Marling made the setting up of the roof-tree and the fitting of the frame the excuse for a regular old-time " frolic " or " bee," as it would be called in the countryside. A particularly good dinner was served in the middle of the day, and in the evening, when the work was finished, they were allowed to dance and sing to their hearts' content, so that they were all made to feel in high good-humour.

The fine bright days continued, and 'the house on deck grew rapidly. The roof and sides being completed, the next business was the packing; for, of course, no ordinary wooden walls would be of much avail against the fearful cold that would come in due time. The securing of the material for packing afforded another pleasant bit of variety. The material was the dry moss which thickly clothed the sheltered portions of the land near by, and which they procured by going on shore with gunny sacks, which were quickly filled with the spongy stuff.

Harold thought it very fine fun at first clambering over the rocks and up among the cliffs, but after a while it ceased to be amusing, and then he betook himself to exploration. He asked his father to allow Patsy to accompany him, and having promised not to go any distance inland, but to keep within gunshot of the shore, he took his gun and a supply of ammunition and started off in great spirits. He felt himself to be a real explorer.

"Why, Patsy," said he proudly, as having scaled the first range of cliffs they looked down upon the great fiord on which the *Narwhal* lay motionless, "perhaps nobody before ever stood where we're standing now; just think of that! S'pose we call this point Harold's Hill, and that one over there Patsy's Peak."

Patsy laughed at the idea of any place being called after him, but Harold insisted that as explorers they had the same right to call places after themselves as other people had; so the hill and the peak were duly christened, and the two boys proceeded to further discoveries.

"Hi, Patsy! look here! What's that?" cried Harold presently, pointing excitedly with his finger to a level stretch of land a mile or more distant, upon which some strange-looking animals could be seen moving slowly. "What queer-looking creatures they are! their heads seem bigger than their bodies."

Shading his eyes with his hand, Patsy looked long and hard, but could make nothing of them. "Sure," said he, "if their heads weren't so big I'd say they were cows."

"I've got it!" cried Harold. "They're moose; that's what they are."

Harold had never seen a live moose, but moose heads stuffed and mounted were often on exhibition at the furriers' in Halifax, and he was quite familiar with their appearance.

"Right you are, sir," said Patsy. "They're moose, and no mistake. What a fine thing it'd be if we could shoot one now! Wouldn't the captain think a deal of us?"

The very same thoughts were running through Harold's mind, and the temptation to go off in pursuit of the supposed moose was very great. But it did not cause him to forget his father's injunction, and his tone was decided enough as he answered,—

"No, no, Patsy; we mustn't do that. The moose are at least a mile away, and we're as far from the ship now as we ought to be. We'd better be getting back as it is."

On their return to the steamer they at once announced their discovery, and Captain Marling had a good laugh at them. "Moose, do you say, Harold? Not a bit of it. No moose in this country. It was reindeer you saw. And just so soon as this job is finished we'll go and hunt them up, I promise you."

Harold shouted with delight at the prospect of a reindeer-hunt, and found it hard to restrain his impatience during the next few days, while the deck-house was being completed. Captain Marling took

good care that this was most carefully done. Upon
the first roof a layer of moss, at least a foot thick,
was laid, and stamped firmly down. Then over this
a second roof of heavy planking was built, and the
cracks between the boards calked with the same
useful moss. This having been satisfactorily accomplished, it only remained to fit the windows and doors
into their places, and the deck-house was complete.

The result was the transformation of the deck into
a very roomy and comfortable chamber, which, although perhaps rather dark in the daytime—for the
windows were necessarily small, and, moreover, had
triple sashes—looked well enough at night, when half-
a-dozen big lanterns shed a generous light into every
corner. Harold highly approved of it. There was
lots of room for the sky-larking in which he loved to
indulge with Patsy, and the very idea of turning the
steamer into a house was full of romance and charm
to him. He had no premonition of how desperately
weary he would become at what now seemed so
delightful in the long, dark, cold days that were
approaching, and with what joy he would hail the
removal of the last plank in that deck-house, which,
being now complete, rendered the *Narwhal* ready for
the winter.

CHAPTER XVII.

A REINDEER-HUNT.

THE preparations for winter were not completed a day too soon. Fortunately, neither were they a day too late. The first day of November saw the steamer's dock, which it had been decided was henceforth to be known as Narwhal Inlet, covered thick with ice; and each day thereafter the shore ice crept further out into the bay, until by the end of the week not a speck of open water was visible from the crow's nest.

"The steamer's in bed now, Harold," said Captain Marling, as, walking out upon the ice, now strong enough to bear a regiment, they looked back at the *Narwhal*, which presented a very snug, comfortable appearance, with her covered decks and shortened masts. "She only needs her bed-clothes, and then she won't mind the cold an atom."

"Her bed-clothes, father! Why, what do you mean?" queried Harold in surprise.

"The snow, my boy, the snow," answered the captain. "When the winter snow comes, we'll pile it up on the roof and around the sides, until she is almost buried in it. She'll be a vast deal warmer for that, I can tell you."

"Then we'll be living in a snow house, just like the Esquimaux, won't we, father?" exclaimed Harold, his face kindling at the idea.

"That's so, Harold; only that our snow house will have a wooden inside, which will make all the difference in the world, as you'll soon see for yourself when we come across some Huskies."

"And when will we see some Huskies, father?" asked Harold eagerly.

"Oh, they'll be along fast enough if there are any of them about," answered the captain. "They'll see our smoke, and come to find out what it means. The sooner they come the better, for I want a couple of them to be our guides when we go after the reindeer. It wouldn't be wise for us to go off entirely on our own hook."

Captain Marling was right in judging that the Esquimaux would come along so soon as the presence of the *Narwhal* was discovered; for the very next day a procession, that Harold thought more interesting than a circus parade, was seen slowly approaching the ship, coming from the south. There were some half-dozen sledges, each drawn by as many dogs, and loaded with an extraordinary assortment of furs, frozen seal meat, household implements, old women and young children, while beside them walked · a number of men, and more women, more children, and more dogs. They were the oddest-looking lot of people Harold had ever beheld; and as they drew near the ship, they could be seen chattering vigorously to each other, and pointing at the steamer, evidently feeling somewhat uncertain as to what this strange-looking thing was, and what kind of people inhabited it.

They were not long left in doubt. The half-breed interpreter whose services Captain Marling had secured at Nachvak now showed his value. Coming forward to meet the procession, he at once entered into conversation with the leaders, who were evidently highly pleased at seeing one who looked and spoke so like themselves. There was, of course, some difference in the dialect, but the interpreter found that he could make himself tolerably well understood, and at once, on behalf of Captain Marling, offered the Esquimaux a hearty welcome.

They seemed at first as timid as they were curious, and having never seen a steamer before, required some coaxing ere they would consent to go on board. But Lane, the interpreter, soon gained their confidence, and placing themselves in his hands, the men followed him up on to the deck, leaving the women and children in charge of the sledges. The number of the men was twelve, and they were the dumpiest, dirtiest, best-natured-looking specimens of humanity that Harold had ever seen. He was exceeding pleased to find that he was actually taller than the largest of them.

"See here, Patsy," he cried to his friend: "I'm as big as any of them."

"So you are, Master Harold, one way," replied Patsy. "You're as high as the best of them, but you're not as broad. Sure, they're as round as a barrel, aren't they, now?"

"Nobody wants to be as broad as they are," said Harold, "or to have such a flat nose, either," he added with a laugh, looking full into the face of a Huskie beside him; who, noticing the laugh, but of course understanding nothing of the boy's uncomplimentary words, returned the laugh with interest, revealing two rows of sturdy teeth as yellow as parchment. His good nature quite won Harold's heart; and remembering his experience at Nachvak, he at once ran off to the steward for a plate of biscuits, which he proceeded to distribute among the visitors, whose bright eyes fairly snapped with delight, for there is nothing they crave more than farinaceous food.

Having permitted his dusky visitors to wander about the deck for an hour or more, inspecting its wonders with childish curiosity, and peppering Lane with excited questions, Captain Marling gave each a bountiful supply of hard tack, and sent them away, first taking care to make arrangements for the reindeer-hunt on the following day. They went away very obediently, and rejoining their wives, who had all the time been waiting for them on the ice with admirable patience, returned to the shore, where they set up their tents, and encamped for the night.

The following day was as fine as heart could wish, and the whole vessel rang with the bustle of preparation for the hunting-party. Captain Marling had determined to give as many of his men as wished a day on shore, for they had now been a long time cooped up in the vessel ; so it was announced that with the exception of the first mate, the steward, the cook, and a couple of sailors, the rest of the ship's company might get ready for a day's outing. The men were highly pleased at this, and with great energy set about equipping themselves. They were allowed entire freedom in the matter, and the consequence was that some of them presented a very ludicrous appearance when they considered themselves ready for the hunt. None of them had rifles, but being bidden by the captain to take any weapon from the steamer's armoury they chose, one man picked up a lance, another a harpoon, a third a flensing-knife, a fourth a blubber-spade, a fifth a chopper, and so on until every one of them had an offensive weapon of some kind.

"If Jack Falstaff could only see them," remarked the surgeon, who had a great turn for Shakespeare, "he would not rest content until he had enlisted every man jack of them in his famous regiment."

However, if they were not very appropriately armed, and had not enjoyed much previous experience in hunting reindeer, they made up in strength and spirits what they otherwise lacked, and nobody was in better humour than Big Alec, who, ever since

Captain Marling's magnanimous action at the time of the mutiny, had been the best-behaved and most valuable sailor on board the ship. Before the day would close he was to have an opportunity of showing by still stronger proof how deep was his gratitude for his captain's forbearance.

The hunting-party, as it moved away from the steamer, presented quite an imposing appearance. It numbered forty, including Harold and Patsy, and at the Esquimau camp it was made up to the even fifty by the addition of all the able-bodied men, who, looking very happy at the prospect of getting a lot of venison when the deer are at their best, were to act as guides for their white friends.

The two boys were in great "fettle." Harold had his own rifle, and Patsy a small one that the captain had loaned him, and which, considering how little he knew about using it, was likely to be more dangerous to himself than to the game. But this view of the case, of course, never entered his mind, and proudly shouldering the rifle, as he had seen the soldiers do in Halifax, he marched along at Harold's side, the happiest boy on earth.

"Won't it be splendid if we each shoot a deer, Patsy?" said Harold. "We might do it, you know."

"Indade, that we might, Master Harold," replied Patsy, full of faith in both Harold and himself. "Sure, this rifle'll kill anything it hits."

They had much conversation of the same kind as, led by the Esquimaux, the party made their way up through one of the valleys, and thence to the level land beyond the cliffs where the boys had seen the reindeer a few days before. There were no deer in sight at first, and Captain Marling divided his little army into groups of five, putting each group under the guidance of a native, and then directing them to spread out in such a way as to sweep the whole plain, and, by describing a sort of semicircle, come all together again at the farther side. About the middle of the day he gave orders for a general advance along the whole line.

The going was pretty rough and wearisome, as the two boys, sadly out of practice in tramping owing to long confinement on shipboard, soon found out. The plain was strewn with boulders, intersected by gullies, and tossed up and down in hills and hollows that made walking anything but an amusement. Captain Marling and Dr. Linton, with Harold and Patsy, composed one of the groups, their guide being a sturdy little Huskie, whose stubby legs carried him over the ground at a surprising pace. By the time they had gone a couple of miles, Harold began to lag, and his father, noticing it, said,—

"Look here, Harold : there's no need of your tiring yourself out keeping up with us. Suppose you and Patsy stay here, and we'll come back and pick you up after a while. The reindeer may turn this way when they're started, and then you'll have a chance at them."

Harold's pride had prevented him from proposing this plan himself, but when his father suggested it he readily assented ; so the three men went on ahead, leaving the two boys at the base of an elevation, which they at once ascended in order to follow the others' movements.

"Now if the reindeer will only be good enough to come right along," said Harold, stretching himself at his ease on the summit of the knoll, "I shall be most happy to have a shot at them."

"It's not many shots we'll be gettin' if we wait for the deer to come to us," said Patsy. "It's *we* must go after them."

The inequalities of the country soon hid the rest of the party from view, and the boys were apparently the only living objects within the bounds of the horizon. After a while they began to feel somewhat lonely, and coming down from their post of observation, set out to follow in the track of the others. When they had gone some distance, they were glad to see Big Alec coming towards them. It seemed that he, like themselves, had found the walking very troublesome, and not having a rifle, but only a whaling-lance for a weapon, had decided to give up the chase of the deer, and wander about as he pleased until the hunt was over. Big Alec seemed no less glad to see them, and the three strolled along together at a leisurely pace.

Presently their attention was aroused by the sound of rifle shots so distant that they sounded more like the popping of corks than the explosion of death-dealing gunpowder.

"Aha! that means business," exclaimed Alec. "I hope they'll drive the deer this way. I'd like to have a look at them."

To judge from the reports of the rifles, the sailor's wish was likely to be gratified, for the firing grew nearer, showing that the deer were coming in their direction.

"Get your gun ready, Master Harold," said Alec, grasping his spear tightly. "They must be coming this way."

Harold made sure that the cartridge was in its place, and then, trembling with excitement, awaited the appearance of the reindeer. They were not kept long in suspense. From the other side of a ridge not half-a-mile away a herd of deer, probably the same the boys had seen before, dashed into sight with half-a-dozen Esquimau dogs barking madly at their heels. They were heading directly for the boys, and when they disappeared in a gully, Harold could not breathe for excitement until they rose into view again.

On they came at a lumbering but swift gallop, headed by a splendid buck with antlers branching out like a forest tree. There were at least twenty of them, and all in fine condition.

"I'm going to try for him," cried Harold, pointing at the leader.

So intent were the deer in escaping from the dogs that they did not notice the boys and their companion until they were almost upon them. Then with a fierce snort the big buck threw up his head, and at that moment Harold, who was standing almost in front of him, levelled his rifle and fired. Down went the buck with such suddenness as to nearly turn a somersault, and off darted the herd, swerving swiftly to the left, just giving Patsy time to send a bullet into the last one, which broke its leg, and rendered it an easy victim to the dogs, who soon pulled it down.

With a glad hurrah Harold rushed forward to possess himself of his prize. But he reckoned without his host. The deer was only stunned, not killed. Just as the boy was within a few steps it sprang to its feet, and furious with pain from the bullet, which had struck it at the base of the horns, charged straight upon him.

"My God! the boy!" cried Big Alec, as he saw Harold's danger.

Harold saw it too, and tried to avoid it. As he did so his foot turned upon a loose stone, and he fell headlong. It was a most fortunate accident. The maddened deer was almost upon him. Another moment and he would have been caught in the tremendous antlers. But the sudden fall saved him, and, unable to check itself, the fierce creature bounded over his prostrate body, giving him more than one sharp blow with its hard hoofs as it passed.

Before he could rise it had turned, and was making for him again. Now was Big Alec's opportunity. With a shout that sounded more like the roar of a bull than the utterance of a human throat, he sprang in front of Harold, brandishing the lance. But what did the reindeer care for whale-lances! They had no terrors for him. Not for an instant did he pause. The great antlers were lowered to receive the lance, knocking it at once out of the sailor's grasp, and then, as he refused to budge, they crashed into his broad breast.

Big Alec gave vent to another roar, this time of agony, for he was sore hurt by the sharp tines, and throwing his arms around the horns, sought to drag the deer to earth. A tremendous struggle ensued. The deer was one of the largest of its kind, and the sailor one of the largest of *his* kind. They were well matched, and both had their fury aroused to the highest pitch. Scrambling to his feet, Harold re-possessed himself of his rifle, and stood at one side watching this extraordinary wrestling-match with intense anxiety. He did not dare to fire, for so rapid and confused were the movements of the combatants that the chances were he would hit Alec instead of the buck.

In the meantime, Patsy, in whose composition the element of fear seemed somehow to have been omitted, had snatched up the lance, and was giving the deer vicious prods in the side and neck, from which the blood was streaming. Yet the powerful creature was clearly getting the better of his opponent. If the sailor could only have got the animal around the neck, he would have been more than a match for it; but his hold upon the huge horns gave him very little advantage, and he was fast becoming exhausted in his efforts to save himself from being gored by the infuriated animal.

"For God's sake, shoot the brute!" he gasped out, finding himself in such straits. Just as he spoke, the buck presented his side broadly to Harold, as the latter stood by watching eagerly for a chance to render his companion aid.

"Now, my boy, now!" cried Big Alec.

Harold raised his rifle, and without waiting to take aim, fired at the buck. The bullet entered just behind the shoulder. The creature gave a wild bound into the air, dragging the sailor clear off his feet, and then fell upon him dead, beyond all doubt this time. The bullet had cleft his brave heart.

Instantly the boys sprang forward to extricate Big Alec from his critical position. With a great effort they dragged the dead deer off; and then, to their great alarm, discovered that he was insensible. They tried such means as suggested themselves at the moment to rouse him, but still, to their great dismay he remained immovable, and as white as he was still.

"Mercy on us!" cried Harold. "Surely he can't be killed!"

If he were not, he certainly looked startlingly like it. His clothes had been torn to tatters in the terrible struggle; the sharp tines of the antlers had wounded him in places upon both his breast and face, from which the blood flowed freely; and in that last desperate bound, the noble buck's death-throe, he had been hurled violently to the ground, his head coming into contact with one of the boulders that lay everywhere about.

"O Patsy, what shall we do?" said Harold, with a sob; for he really feared that the big fellow, who had come in so bravely between him and danger, had forfeited his life. He would have had his sympathy drawn out for any one in such a situation. It was intensely strong for one who had risked so much for him.

"Sure, sir, he can't be dead; he's only fainted," replied Patsy, in a tone that betrayed his own anxiety. "Let's get the blood off his face, and he'll come to all right."

There was a pool of water near by, and dipping his

*"Throwing his arms around the horns, big Alec sought
to drag the deer to earth."*

Page 246.

handkerchief in this, Harold gently bathed Big Alec's face, being rejoiced to find that it was not so badly cut after all, and that the blood was easily stanched. Still, Alec remained motionless, while from the more serious wounds on his breast the life-blood was slowly ebbing away.

"Oh, why don't they come!" cried Harold, wild with anxiety and the sense of his own helplessness. "If Dr. Linton were only here! Do go, Patsy, and see if you can see anything of them."

Patsy at once ran off, and Harold was left alone with the insensible sailor, whose life now hung upon the chance of Dr. Linton reaching him within a very few minutes. 'Poor Big Alec! he was drawing very near death. Each moment saw his chances diminishing, and if anything was to help him, it must come speedily.

CHAPTER XVIII.

ESQUIMAU EXPERIENCES.

WHEN each minute seems an hour, one cannot measure accurately the flight of time, and Harold could not tell how long Patsy had been gone, ere he returned, running at full speed, and so out of breath that he could hardly say the words, "Dr.—Linton's—coming—just—behind!" before the doctor himself appeared, and at once threw himself down beside the still motionless body. He felt Big Alec's pulse, and then placed his hand upon his heart.

"Not dead yet!" he reported, looking very much relieved. "But we must stop this blood right off. Get me some moss, quick, boys." The boys hastily gathered up some handfuls of moss and handed them to him. "Now, your handkerchiefs." Having got the handkerchiefs, the surgeon with deft hands proceeded swiftly to make bandages for the wounds, and soon had the bleeding completely under control.

"So far so good," said he. "Now, to wake him up."

Patsy brought his hat full of water from the pool, and Big Alec's face was drenched with it, Harold meantime chafing his hands vigorously. This treatment was soon successful. The sailor stirred, opened his eyes, closed them again, and then put up his hand to feel his head, muttering, in a bewildered way, "What's the matter? Who struck me?"

In a few minutes more he had quite regained his senses, and made an effort to get up on his feet, but Dr. Linton restrained him. "No, no, Alec," he said kindly; "you must lie there until we can carry you to the ship. If you try to move, you'll start your wounds bleeding again."

"As you say, sir," answered Alec. "I do feel pretty well shaken up, and I'll just keep still."

A couple of the sailors were then despatched to the steamer to procure a hammock and a pair of hand-spikes, with which a sort of stretcher was improvised, and four strong seamen bore the wounded man to the *Narwhal*, where, by Captain Marling's direction, he was placed in Collins's state-room, as he would be better off there than in the dark and stuffy fore-castle.

If any feeling of resentment had lingered in the captain's mind, it was completely and finally banished when he heard of Big Alec's bravery in the boy's behalf. He lost no time in expressing his gratitude, and in assuring the well-pleased sailor that he had by his gallant action secured a friend whose memory of that service would not be short-lived.

The hunt had been highly successful, no less than twelve fine fat bucks and does having fallen victims. These were equally divided between the Esquimaux and the steamer, and if ever people looked happy, it was those dusky, dumpy savages, as they toiled home to their tents, bearing their heavy loads of delicious venison; for, as a matter of fact, they have no higher ideal of bliss than a stomach filled to repletion with favourite food.

They had a glorious blow-out that night—men, women, and children eating and eating and eating, until it seemed a marvel they did not share the fate of the frog who strove to be as big as the ox. But then it was an era in their lives to have an unlimited supply of reindeer meat; for, having no guns, it was not often they were successful in hunting these fleet-footed creatures. The white visitors' brilliant success as hunters, and their generosity in dividing the spoils, quite decided the Huskies to make their winter quarters in close proximity to the steamer, of which plan Captain Marling highly approved, as he wished to study this interesting people closely; and, moreover, the band was too small to be much of a burden, even if he had often to contribute to its support during the long winter.

Harold and Patsy were delighted at the idea of having such quaint neighbours, and promised themselves fine fun learning how to manage an Esquimau dog-sledge, and in turn teaching the Esquimau boys how to handle a gun. They were disappointed at first because the natives lived in tents of dried reindeer-skin instead of snow houses; but the captain explained that the tents were only summer residences, and that the snow houses would be built so soon as the firm, dry winter snow had come.

Sure enough, after the first heavy snowfall, and when there was no longer any doubt but that winter had come to stay, the Esquimaux set about making preparations for their winter abode; and the two boys, who now spent the greater part of every day ashore with the natives, watched them with eager interest. Their first proceeding was to hunt out a satisfactory bed of snow, which they did by means

of snow-testers—long, thin rods of reindeer bone—
which they thrust through the crust down into the
snow beneath, to make sure that it was well packed,
and fit for their purpose. A good bed having been
found not far away, they at once set to work to
build their "igloos."

Of many a snow house had Harold been the proud
architect, but he felt that his most ambitious efforts
would look little better than ant-heaps beside the
symmetrical structures these ignorant natives built
up like magic before his eyes. Taking a long,
strongly-made knife, the Esquimaux would describe
a circle in the snow of about ten feet in diameter,
thereby indicating where the base course of blocks
was to be laid. Then the blocks were cut out from
the bed, and laid around this circle. These blocks
were the size of a large pillow, and weighed about
twelve pounds apiece. They were laid upon their
edges, not like bricks in a series of courses, but in
one spiral course that ascended without break from
foundation to summit, changing in the ascent from
a rectangular to a triangular shape, the cap of the
dome being formed by three triangular blocks meeting in a key block that held all firmly together.

Harold was filled with admiration for the skill

with which the dusky builders fashioned the blocks
and fitted them together. Borrowing one of the
sailors' sheath knives, he tried his own hand at it,
but soon gave up in despair, for the blocks that he cut
would not do at all. They were of all shapes and
sizes, and utterly refused to lie snugly together.

"I'll have to get one of the Huskies to teach me,
I see," said he, throwing down his knife; "but I'm
bound to learn."

When the dome had been completed, a hole was
cut at one side for a door, and a long covered way
built to keep the wind from blowing in too freely.
This covered way was, of course, very low, and one
had to go through on his hands and knees. The next
business was to chink the crevices between the blocks,
which was done by cutting off a little snow from the
edges of the blocks and ramming it into the cracks
and crevices with a blow of the fist. Finally, a foot
or two of loose snow was heaped over all the dome
except at the entrance, and the "igloos" were completed so far as the outside part of them was concerned. Then came the furnishing of the inside, and
if the boys had been filled with admiration before,
they were overflowing with wonder now. These
clever workmen in snow proceeded to make a plat-

form which was about thirty inches high, and took up nearly three-fourths of the inside space. Upon this they spread a layer of moss and the reindeer-skins, and in response to Harold's wondering inquiry showed very clearly that this platform was intended for their bed.

The boys were fairly staggered. A snow house seemed tolerable enough if you could get nothing better; but a *snow bed!* that was quite too much of a good thing.

"Did you ever hear the likes of it?" exclaimed Patsy, when he understood the matter. "Sleeping in a snow bunk! Sure, I'd be frozen stiff the first time I tried it."

When, later on, after the thermometer had got down away below zero, and the keen air cut like a razor, the boys visited one of these "igloos," and not only saw its inhabitants lolling about comfortably in a temperature that did not melt their snow beds, but beheld two chubby, dirty children, *stark naked*, rolling over one another on the reindeer-skins like a pair of frolicsome kittens, they came to the conclusion that the Esquimaux were made of very different stuff from themselves, and that they certainly could never get used to their style of living. Just imagine what it would be like to go through a whole winter without ever feeling really warm. Talk about "perishing with the cold!" The people in the temperate zone don't know what it really means.

Alongside the large "igloo" a smaller one was built, to serve as a sort of pantry or store-room, and in this the reindeer meat, the seal blubber, and the other supplies were kept, as well as the harness for the dogs, which could not be trusted within reach of those omnivorous creatures. Harold was very much disappointed in the Esquimau dogs. He had expected them to be fine large creatures, something like the Newfoundland dogs; but instead of that they were shaggy-haired, sharp-nosed, wolf-like animals, about the size of a collie, with nothing attractive in their character, being almost as wild as wolves, and having no other idea of obedience than cowering before the crack of their master's whip. Their only redeeming feature, next to their ability to drag a sledge, seemed to be the simplicity of their appetite. They would eat anything and everything, and Harold used to amuse himself experimenting upon them. One of them that a sailor succeeded in coaxing on board the steamer showed his appreciation of Jack's hospitality by devouring a cloth hat, a boot, the best part of a flannel shirt, and one leg of a pair of trousers before he was detected in his mischief.

When the natives completed their "igloos," which they built in a sort of circle, their little village presented quite a snug and cozy appearance from the outside, whatever thin-skinned, cold-blooded writers from the South might think of the inside, and it was evident that they considered themselves particularly fortunate in the selection of a site for their winter settlement; for they were the very picture of contentment, as they waddled about in their furry garments from igloo to igloo, or from the settlement to the steamer. There was a constant interchange of visits between these neighbours. The white men went to the natives to study their habits and mode of life, and to try to pick up something of their language. The natives came to the white men to study their bread and their beef, and to try to pick up any unconsidered trifles that might be lying around handy. Not that they were light-fingered and unworthy of trust. On the contrary, they would not appropriate so much as a pin' without first asking permission. But then they were no more bashful about asking than children in short frocks would be. Harold, with that easy benevolence characteristic of those to whom the giving costs nothing, was at first disposed to grant so many of their requests that his father had to interpose, and order him to give nothing more away without first asking him.

Not a day passed that, alone or accompanied by Patsy, Harold did not pay a visit to the igloos. With the help of Lane, the interpreter, he began to master the "Innuit" language, and would proudly repeat to his father every new word he acquired. One of the first sentences he got a good grip upon was the curious but eminently appropriate way in which the Huskies say good-bye—namely, "*Ta-bourke aperniak in atit*," which in plain English meant, "Good-bye; don't bump your head." Harold thought this exceedingly amusing, and resolved to carry it home with him, though he forgot everything else. As for Patsy, he found he could make himself so well understood by dint of plenty of gestures and grimaces, that he left the language of the tongue alone, saying that his "unruly member" was altogether too clumsy to get round the native speech.

The boys' chief source of delight, however, was being taught how to manage an Esquimau dog-sledge. This was an art by no means easily learned. The dogs were utterly unruly, having got pretty well out of training during the summer; and, moreover, they evidently had great contempt for drivers who could not hurl exclamations at them in Innuit, or wield the long-thonged whip with which their proper masters could take a tuft of hair out of their backs at ten yards' range. The sledges seemed such clumsy, primitive affairs in the boys' eyes, made as they were of roughly-fashioned driftwood lashed together with reindeer-gut, that they had the ship's carpenter build one for themselves after the most approved civilized model, and put together with nails. This they brought out with great pride and showed it to their heathen friends, as much as to say, " Here, you ignorant Huskies; this is something like a sled. Just see how much better it is than your clumsy affairs."

But, much to their surprise, the Esquimaux did not seem at all impressed; on the contrary, they shook their heads in evident disapproval of the new model, saying something that no doubt meant, " No good; our kind the best. White man don't know how to make sledge for Innuit."

As the boys were not to be convinced by words, the natives determined to prove their case by practical illustration. Accordingly, the best team of dogs they had was harnessed carefully to the new sledge, and the best of their drivers being put in charge, Harold and Patsy were bidden to jump on, the long whip cracked like a pistol-shot, and off they went at full speed. The way was fairly level, but very rough, and the sledge bumped along, seeming to strike every obstacle within reach in its wild course. The superiority of gut over nails as a fastening soon became evident. Instead of yielding a little to every shock, the sledge opposed it stiffly, and carefully as it had been put together the incessant bumping strained it apart, until at length a particularly violent concussion with an ice hummock broke one of the runners short off, pitching the boys out on their heads, and causing their Esquimau driver to break into a merry peal of laughter.

Harold and Patsy picked themselves up, rubbing their heads ruefully, and gazing at the shattered sledge with an expression that said plainly enough, " The Huskie's right. He does know his own business best. I guess we can't improve on his way of building a sled."

But if they could not teach the Esquimaux anything regarding sled-building, they certainly thought the native houses were capable of great improvement. However pretty and romantic they looked when freshly built, they soon proved themselves hollow frauds, so far as keeping warm the visitors from the South was concerned. The walls were far from air-tight, the outside air passing through as readily as it would through a lump of white sugar held between a boy's lips; and as this outside air was always well below zero, inside warmth was hardly attainable when the only means of heating consisted in a couple of small stone lamps, in which a feeble flame flickered all day long. Even supposing it had been possible to make sufficient heat to warm up these snow huts properly, their construction put this out of the question. The snow must not be allowed to melt. The temperature inside must, therefore, be kept below freezing-point. In other words, the Esquimaux in their winter homes live in a temperature so chilly that one might almost make ice-cream there without having to put iced salt on the freezer. And yet, as Harold noticed, they never seemed cold, and their children would play about without clothing in this freezing atmosphere.

When the whole family gathered in the igloos, and the stone lamps were burning at their feet, thawing out blubber or venison for dinner, the heat ascending to the roof would begin to melt the points and edges

of the blocks above. Then somebody would take a
handful of snow from the floor and paste it on the
leak. Then if the heat continued these snowballs
would become saturated with water, and— But let
Patsy's experience illustrate the consequence.

Harold and he were sitting in an igloo one cold
day, and out of respect for the feelings of their
guests, the Esquimaux had warmed the place up as
much as they dared. Indeed, coming in from the
bitter cold the boys found the place almost comfortable.
Presently Patsy, who was lounging on the edge of
the bed with the hood of his fur coat thrown back,
felt a drop of water falling upon his head, and looking up saw that the roof was beginning to melt. One
of the Esquimaux noticed it also, and promptly applied
a snowball to the leak, stopping it at once. The
cause of his discomfort having been removed, Patsy
resumed his place, and was lolling there at his ease,
when suddenly he sprang to his feet with a shout
that startled the others, and putting his hand to the
back of his neck, cried,—

"Oh, musha, musha! what's this that's struck me?
Ouch! there's a snake down my back, sure."

It wasn't a snake, of course, but it was something
almost as bad. The snowball put up to patch the

leak having become saturated with water, had fallen off, and found its way with perfect aim in between Patsy's hood and the back of his neck, the icy slush slipping down his backbone with an effect which may be easily imagined.

Thenceforward Patsy never sat in an igloo without keeping one eye upon the roof and changing his position at the slightest suspicion of a leak. One lesson was enough for him. Indeed, it would have been for any one. The incident, moreover, caused Harold and him to wonder still more how these simple Esquimaux could pass winter after winter in such comfortless abodes. It also made them more grateful for their own snug quarters, and for the far more pleasant homes they hoped by-and-by to see again.

CHAPTER XIX.

AN ARCTIC WINTER.

WINTER had come, and such a winter as no one on board the *Narwhal* had ever experienced before. Down, down, down crept the mercury in the thermometer, until at length it could get no farther, and then one night the registering glass that hung at the door of the deck-house froze solid, so that you could have handled the mercury just as though it were a stick of candy.

Moveover, as the mercury sank the days shortened, the period of daylight growing briefer as the close of the year drew near. The immense value of the deck-house now became evident. For the men to have been confined to the forecastle—which, although more roomy than such places ordinarily are, was still none too spacious for twoscore men—would have been very trying to their temper and spirits. But the huge deck-house, lit by half-a-dozen lamps, and warmed by a

couple of big stoves Captain Marling had not forgotten to provide, made a splendid refuge from the
cold and dark, where the sailors could read, sew, make
curious carvings out of wood, or play games, according
as they pleased. They, of course, had nothing to do
but to keep the ship in order, and although the captain's discipline never relaxed so far as the control of
the men was concerned, they were allowed abundant
freedom of action within proper limits. They were,
upon the whole, a very contented, peaceable lot of men,
and if they ever did feel disposed to grumble at any
of the hardships inevitable in their situation, they had
the sense to keep it to themselves. The tragical consequences of Collins's evil counsel had taught them a
lesson they were in no hurry to forget.

The one thing to be feared by men situated as
were those on board the *Narwhal* was the scurvy.
The constant eating of salt food and the lack of
active employment made this dread disease a contingency to be carefully guarded against. There was not
so much danger of it in the saloon, for the captain
had laid in an abundant supply of canned meats and
vegetables; but the sailors, of course, had the regulation fare of salt junk and hard tack, and if they
were to do nothing but loll about trying to keep them-

selves warm, the probability was that they would be down with the scurvy ere the winter was half through.

Here, again, the captain's forethought found illustration. He had anticipated all this, and had made provision against the difficulty in a way that did credit to his ingenuity. Calling the men together one fine morning when the thermometer was about twenty-five degrees below zero, he asked them how they would like to have a game of base-ball. They received his question in much the same way as if he had asked how they would like a slice out of the moon. But he assured them he was in earnest, producing a base-ball as a token of his sincerity. He then explained his scheme, which was simply that a space about a hundred yards square should be cleared of loose snow, and thus converted into a ball-field the like of which certainly did not exist anywhere else on the continent. With big lumps of coal, which would show out plainly on the white ground for bases, and handspikes for bats, there was nothing to prevent their having lots of fun, even though the game in skill and style fell far below the League standard.

The men took hold of the idea at once. The space required was carefully cleared and smoothed, and thenceforward every day that the weather permitted the most of them were to be seen playing base-ball to the very best of their ability.

The scene was a most curious one, and ludicrous in the extreme. Every player was little better than a moving mountain of furs, some of them being muffled up until only their eyes and nose were visible. Their hands were encased in mits thicker than any catcher's gloves, and a good fly catch was almost an impossibility, although now and then some lucky fellow would, quite as much to his own surprise as to that of his companions, succeed in performing the difficult feat. Tumbles were the rule rather than the exception, and sliding for bases was the regular way of getting there. Captain Marling generally filled the important position of umpire, doing so with a zeal and gravity worthy of a League official.

Harold and Patsy were, of course, in ecstasies over the base-ball. No game was complete without them, and they were looked upon as the "mascots" for the respective sides, a very active but good-natured rivalry being created. First one side and then the other would develop a "wonderful batting streak," and the scores were kept with great accuracy, so that comparisons could be made at the end of the week, and the averages made up. It need hardly be said that there were no games postponed on account of rain while the season lasted. So that an Arctic base-ball field proved itself to be not entirely without advantages.

Although the only thing in the way of a grand stand was a big snowdrift at one side of the "diamond," the game was not without spectators. The Esquimaux were always on hand, laughing merrily at the mishaps of the players, even though they could not appreciate the good points of their play. If the ball chanced to roll their way, they would all make a rush for it, each trying to be the first to pick it up and return it to the nearest player.

By-and-by the sun went down behind the horizon, and did not come up again. The long Arctic night was upon the people of the *Narwhal*, and for three weary months they were doomed to constant twilight— the Aurora Borealis, that flamed and flashed across the face of the Northern firmament with an indescribable variety of splendour, being their beautiful but insufficient substitute for the sunlight.

Harold did not take kindly to the idea of parting with the sun. It seemed very strange to wake up in the morning and find it no brighter than it was when he lay down to sleep; and although it was not dark enough to prevent one from going abroad as usual, still the range of vision was very limited, and constant care had to be exercised.

Base-ball, of course, became no longer possible. The sharpest eye could not follow the flight of the sphere in that uncertain light; and the men were lamenting their hard fate, when again the clever captain proved equal to the occasion. He had foreseen all this when he was making ready for a winter in the land of darkness; and from that mysterious locker, into which not even Harold was permitted to peep, now produced a ball which, under Dr. Linton's manipulation, assumed the proportions of a pumpkin.

"Here, Harold," said the captain, tossing him the huge sphere. "If you can't see a base-ball in this dim light, you'll have no trouble in seeing this. We'll play football after this."

And play football they did, day after day, during the remainder of the winter. In this game the Esquimaux became active sharers. They could run and tumble and trip and pick up a pumpkin as well as any of their white friends, and they entered into the fun with amazing vigour for such fat little fellows.

As the month of December drew toward its close, Harold began to take a deep interest in the almanac,

and to count the days that still remained. Evidently
he and Patsy had something important on their minds,
which would in due time be declared. The week before
the twenty-fifth the matter was made known. Captain Marling was sitting in the saloon reading, when
Harold stepped up quietly, and putting his arm around
his father's neck, said, in a very meaning tone,
" Father, do you know that next Wednesday will be
the twenty-fifth ? "

Captain Marling looked up, and there was a sly
twinkle in his eye as he answered, " I hadn't thought
of it, Hal.   Why do you mention it ? "

" You know well enough, father."

" I'll know better if you tell me, Hal," returned the
captain, who was apparently impervious to hints.

" Why, it's Christmas, of course," said Harold,
despairing of getting his father to say it for him.

" Sure enough, my boy, sure enough !   What a pity
we were not at home, so that we might keep it ! "
And the captain's face grew contemplative.

" But we will keep Christmas, all the same, won't
we, father ? " exclaimed Harold, somewhat anxiously.

" Keep Christmas up here !   And pray, sir, how
shall we manage that ?   No woods to get a Christmas-
tree in ; no shops to buy Christmas presents in."

And the captain smiled, as though to imply that Harold was talking nonsense.

But the boy was not so easily rebuffed. He more than half suspected his father of being only in fun; and, any way, he intended to have just as good a Christmas as could be managed under the circumstances.

"If we can't get a Christmas-tree or buy presents, we can have lots of fun, all the same, father," said Harold, in a very determined tone; "and I mean we shall have it, if you have no objections."

"Not a bit, my boy, not a bit," laughed the captain. "Go right ahead. I'll give you full charge, and will do my best to help you. Just ask for what you want, and you shall have it."

Harold sprang upon his father, and gave him a hearty hug. "You dear, good father! You're just the best father in the world. You see if we don't have a good time."

For the next few days Harold was full of business. He made Patsy his aid-de-camp; and the two boys were always consulting together with an air of great mystery and importance. The captain gave him *carte blanche* as regarded the stores, and instructed the cook and steward to assist him to their utmost.

The boys' plans were simple enough, but the execution of them afforded them a great deal of happiness. The programme was as follows:—There would be divine service in the morning, because when at home he had always gone to church on Christmas. This his father would conduct, just as he did the service every Sunday. In the afternoon there would be a grand football-match, in which everybody on board that could must take part, to be followed by a number of races. And then, in the evening, a big dinner in the deck-house, which would be decorated with flags and pictures and furs, to the extent of the full resources of the ship.

Christmas came, and everybody was ready to celebrate it heartily. Harold and Patsy were in high glee, and could scarcely maintain due decorum during service, so that it was quite a relief to them when the football began and they could give free vent to their feelings by shouting and running to their hearts' content. The match was a great success. Twenty players took part on each side, and after a very exciting contest and innumerable ludicrous incidents, Harold's side came off victorious by three goals to two for Patsy's side. The races went off equally well; and then, with appetites worthy of Arctic wolves, they all returned to the ship to make ready for the dinner.

The deck-house had been decorated under Dr. Linton's direction, until it bore quite a cozy and home-like appearance. Two long tables had been laid down the centre, and spread with the captain's best napery, cutlery, and glassware. The saloon chairs and lounges were brought up, and everything done to make a fine appearance. From the captain to the cook, everybody donned his best suit; and thus attired the men of the *Narwhal* made up a company that any captain in the world might have been proud to command, as they took their seats at the table with watery mouths.

The dinner was a veritable triumph. With Christmas in mind, the far-seeing captain had directed two of the finest haunches of reindeer venison to be put away, frozen, at the time of the hunt; and these, roasted to a turn, now adorned the heads of the tables. Besides that, he had permitted Harold to raid his supplies of canned meats, and delicious duck, toothsome turkey, marrowy tongue, and other dainties rare to the sailor's palate gave out a fascinating fragrance, while sweet corn, green peas, and red tomatoes helped to make up a feast which certainly had never been surpassed in that latitude.

The sailors were in the seventh heaven of enjoyment. The captain and his party were their hosts, and they saw to it that no man lacked for anything. Harold, as entertainer-in-chief—for his father would have it that this was *his* dinner—was never still, going about from man to man, and pressing upon each what he thought would be liked best.

At the close of the dinner Captain Marling made a short speech, referring to the associations of the day, and then the health of their gracious ruler and that of the dear folks at home was remembered with many a cheer.

When they all were quiet once more, the captain stood up and announced that the second part of the day's entertainment would now take place. Everybody knew what was coming, for Harold had made no secret of the matter. But they were none the less eager, notwithstanding.

Some fifteen minutes previously Patsy had been sent as a messenger to the Esquimaux, bearing an invitation for the entire party to come over in a body to the steamer, where there was something pleasant in store for them. They accepted, of course; and now the men, women, and children, to the number of fifty at least, were pressing into the brilliantly-illuminated room, whose radiance almost blinded their eyes, unaccustomed to such light. Harold gave them a warm welcome, and led them to the after-part of the room, where, in the open space, a large object, shrouded in canvas, had been looking very mysterious. With happy, expectant faces they crowded about him, while the crew surrounded them with a ring of interested spectators.

Harold clapped his hands as a signal for somebody unseen. The canvas cover was suddenly whisked away, and lo, before the gaze of all stood a veritable Christmas-tree, covered with all sorts of decorations and parcels, and dotted with candles, which the two boys proceeded rapidly to light. There was a vigorous round of applause from the ship's company, in which the Esquimaux, not understanding anything more than that everybody was in high good-humour, joined by grinning to the full extent of their capacious mouths.

At first sight, and viewed from a little distance, the Christmas-tree looked natural enough; but on closer examination it would be seen to be such a tree as probably had never been set up in honour of

Santa Claus before. There were, of course, no trees to be had far north, and Harold would have been compelled to do without what seemed to him the chiefest part of a Christmas celebration, had not his friend Lewis come to his aid.

"I'll make you a Christmas-tree, Hal," said he, when Harold had told him his difficulty.

"How can you do that?" asked Harold eagerly.

"Easy enough," replied Lewis. "Make it out of reindeers' horns."

Harold could not understand it at first, but Lewis soon made his meaning plain. A post was set up as a backbone, and to this the branching antlers were secured in such a way that when the work was complete the effect was capital, and the resemblance to a tree quite striking.

Harold was delighted with it. He thought it even better than the regulation spruce cone; and now that it was laden with gifts for the people of the very region from which old Santa Claus is popularly supposed to set out on his annual joy-giving journey, it seemed to him the finest Christmas-tree in his experience. The gifts were both numerous and appropriate. Everybody on board had contributed something. There were knives and hatchets for the men, needles and beads for the women, cakes and sweetmeats for the children, and many another thing besides; while piled about the foot of the tree were a number of bags, one for each household, containing biscuits, sugar, tobacco, and salt—delicacies as rare among the Esquimaux as are canvas back and terrapin among ordinary folk to the south.

The scene in the deck-house that evening was an exceedingly happy one, and Harold vowed that he had never spent a more delightful Christmas in his life, thereby showing very clearly how much more blessed it is to give than to receive. Having been well feasted with the remains of the dinner, the natives went back to their igloos, not much wiser, perhaps, on the subject of Christmas, but certainly very much pleased with the conduct of their white friends.

The longest winter, like the longest night, must come to an end, and in due time the combination of night and winter the *Narwhal* was experiencing passed away, and the sun returned. Each day he stayed a little longer in the heavens, and each day the effect of his beams became more perceptible. As the days grew longer and warmer the people of the *Narwhal* grew restive. They were utterly weary of their inactive life, and burned to be off on their homeward voyage. Not even Captain Marling was exempt from this restlessness of spirit, and it was in order to pacify it, for a time at least, that he planned an exploring expedition which came very near bringing about a terrible catastrophe. But the story must wait for the following chapter. We may, however, add just here that it led them into a peril grave beyond the others, from which by-and-by they could tell of their rescue—a peril, too, that added greatly to Harold's reasons for giving special thanksgiving to God. He had stored up many before, but this added one that he never forgot. But we must go on to tell of it.

CHAPTER XX.

HOME AGAIN.

THE northern side of Garry Bay was formed by a great stern headland that thrust its barren bulk far out into the waters of Boothia Gulf. At the point it rose into a peak whose summit commanded an immense stretch of land and sea. Growing impatient at the slow advance of spring, for whose warm hand he must needs wait to unlock the icy fetters that bound his ship fast, Captain Marling determined to attempt the ascent of this peak, to see if from so lofty an eyrie he could catch a glimpse of open water in the gulf beyond, for as far as eye could see from the *Narwhal* the ice was still unbroken.

The time was the latter part of April. The days were already moderate in temperature,—that is, they were like ordinary winter days in Halifax,—and there ought to be no difficulty in carrying out the captain's scheme. Harold was very glad to. hear of

The start for the Peak.

it. Nothing would suit him better. "Of course you'll let me go, father," said he confidently.

"Oh, I suppose so, Hal, if you promise to be a very good boy," answered the captain.

"And may Patsy come too, father?" asked Harold; "he'd like to ever so much."

The captain hesitated a moment, and then, as if something had just occurred to his mind, said, "Yes, yes, certainly; you may both come. It will probably be our last picnic."

They started the following day, the party being made up as follows: Captain Marling, Dr. Linton, Frank Lewis, both the engineers, Harold, Patsy, and half-a-dozen sailors with Big Alec at their head. Then there were three Esquimau sledges, heavily laden with tents, provisions, and firewood, under the charge of native drivers, who cracked their long whips and looked very important as the procession moved away from the ship.

Everybody had to walk of course, and the going was none too good, so that the rate of progress was not more than about three miles an hour. This, however, would take them to the base of the mountain at least an hour before the early nightfall, and that would allow them sufficient time to put up their camps, and make themselves comfortable for the night. For the first hour or so the boys kept well up with the head of the party, but after that they began to lag and drop back, until presently they brought up the rear, and more than once the captain had to call out, "Brace up, boys, brace up! there's no time for loitering."

Harold looked very longingly at the sledges, which the well-trained dogs seemed to have no difficulty in dragging over the snow that bothered him so much. He was too proud to confess himself tired out, and would rather have dropped in his tracks than beg for a lift. But oh, how glad he was when his father, looking back and noticing his lagging gait, called out, "Getting played out, Hal? Well, you and Patsy just jump upon the sleds for a while and rest yourselves."

The boys did not need to be told twice. The next moment they were each upon a sledge, and thus, by alternate walking and riding, reached the camp ground in good condition. The tents were pitched in a snug little pocket at the mountain's foot, the fires made, the supper cooked, the dogs fed, and then, rolling themselves in furs, the whole party lay down to sleep as only tired men can sleep. The long night passed quietly away, the only sound that broke the stillness being the occasional barking of a dog given to disturbing dreams, or the snoring of some heavy sleeper.

All were awake at dawn, and glad to get the cup of steaming hot tea the cook soon had ready for them. Then preparations were made for the ascent of the mountain. With the exception of one Esquimau, who was left in charge of the dogs, and two sailors to take care of the tents, the whole party set out. The day was perfect—the sky unclouded, the sun in full force, and the wind light.

" We have everything in our favour," said Captain Marling to Lewis, " and, barring accidents, we ought to be back here early in the afternoon."

" I shouldn't wonder if we need to keep sharp look-out for avalanches," answered Lewis. " The sun seems to have already melted the snow a good deal about here."

" That's so, Lewis," returned the captain. " We must keep our eyes and ears open."

They divided into two parties, Captain Marling taking the lead of one, and Frank Lewis that of the other. The two boys were with Lewis, as the captain wished to go on ahead, leaving the others to follow in the path marked out by him. For a time their progress was not very difficult, the winter's winds having packed the snow firmly about the mountain's base, so that the foot made little impression upon it. But as they climbed higher and higher the ascent became more arduous, and Harold found that it taxed his strength severely to keep his place in the party, although the pauses for rest were frequent.

"Sakes alive, Patsy, but this is hard work!" he exclaimed, throwing himself down upon a snow-bank. "If I wasn't so bound to get to the top of this old mountain, I'd just stay where I am until the others came back."

"Sure, I'd be glad enough to stay here as it is," replied Patsy, who was not at all ambitious for fame as a mountaineer or eager for a fine view.

"Oh no! we must go on to the top," returned Harold, rising to his feet. "Come along with you." And off they started again.

Toilsome and tiring work it was: now creeping carefully on hands and knees around a perilous corner, now climbing in the same fashion straight up some slippery slope, then walking in Indian file along a narrow ledge where a single false step meant instant destruction, there was demanded a constant exercise of watchfulness and care. Each party of climbers was strung out upon a long rope, which they fastened around their waists, thus making it impossible for one to be carried away unless all went together. The two boys were in the centre of the Lewis party, and many a serious tumble would they have had but for the sustaining rope. Onward and upward they toiled as the morning hours slipped away, often halting for a few minutes to take breath and rest their wearied limbs. They found the good of the football practice, for it had kept both wind and muscle in working order. Indeed, but for it the ascent would probably have had to be abandoned as beyond their powers.

After two hours of climbing, they came to a spot where the captain ordered a halt to consider the situation. This was far from promising. Before them stretched a long, smooth slope of snow, which had the rugged mass of the mountain-top above, and below the dark abyss of a gorge whose depth there was no estimating. The only way onward lay across this perilous slope, and Captain Marling hesitated before attempting it. When Lewis came up they consulted together for some minutes, and at last the captain decided to make the venture first with his party, leaving Lewis and his to follow, in event of the passage across proving not so dangerous as it appeared. Accordingly, taking with him one of the engineers, the most active of the seamen, and the sturdiest of the two Esquimaux, and seeing to it that they were securely fastened to the rope, Captain Marling attacked the slope. Each man carried a good ice-pick, and the leader at the critical places cut holes for the feet before proceeding.

"Doesn't it look dangerous, Mr. Lewis?" said Harold, drawing close to the second mate, for he felt very nervous. "Do you think father will get across all right?"

"To be sure he will," answered Lewis cheerfully. "He'll do it in about ten minutes."

Slowly but surely the four men crept onward until they were half-way across. Then they rested a moment, and the captain, looking back, waved his hand triumphantly, and shouted, "It's all right! there's no danger!"

He had better have kept silence. His rash boast echoing from peak to peak through the still air aroused the sleeping demon of the mountain. Far up above him, among the scars that seamed the summit, there was a sound of rustling that swiftly swelled into an ominous roar. The captain heard it, and looked upward with a startled glance. Lewis heard it, and cried out in an agony of alarm, "God help them! the avalanche!" But there was no time for rescue or retreat. With awful speed a vast mass of snow, that had been hanging up there ready for a fall, leaped over the edge and went hurtling down the slope, sweeping all before it into the abyss beyond. For an instant the air was dense with particles of snow; then it cleared, and on the spot where four strong men had the moment before been standing, bracing themselves to withstand the shock, there was not even a mark of their footprints. The mountain had conquered, after all!

Struck dumb with horror, the spectators of this terrible catastrophe stood motionless. Lewis was the first to recover himself.

"All hands to the rescue!" he cried. "There's not a moment to lose."

Without a word the others followed him as he set off on the path they had come, Harold feeling as though it were all some dreadful dream from which in time he must awaken. Lewis's keen eye had caught sight of a ledge leading down into the gorge where Captain Marling and his men had been hurled.

If they could only get to them they might yet be able to save them. The ledge fulfilled its promise; and hastening down as rapidly as they dared, they reached the upper end of the gorge. To Lewis's delight, this gorge proved to be not so deep as it seemed from above; and clambering along over the tumbled snow that filled it half full, they came at length to the very spot where the avalanche, bearing its human prey, had taken its mad plunge. There was no doubt about the place, for one of the ice-picks projected through the snow like a signal of distress.

They had no spades, but the snow was loose, and at it they went with their hands, working for dear life. Presently a glad shout from Lewis announced that something had been found. It was a foot. Digging away furiously, the whole body was soon uncovered, and behold it was the seaman who had brought up the rear. The rope was still attached to him. He was insensible, but breathing.

"Hurrah!" cried. Lewis; "we'll have them all soon."

The digging went on with redoubled vigour. The engineer, the Esquimau, and lastly Captain Marling were disinterred. All were unconscious, but happily none were dead. The shock of the avalanche, the fall over the cliff, and then the smothering plunge into the deep snow had driven them into insensibility; but strange to say, beyond a few scratches in the face, none seemed to have received any other injury.

Under Dr. Linton's direction, vigorous measures were taken to restore the rescued ones. These were speedily successful. One by one they regained consciousness, and then, to the delight of all, it was discovered that not a bone had been broken or serious hurt suffered.

"Let us thank God for our most wonderful escape," said Captain Marling fervently, when he had quite recovered himself. "We will not risk our lives in another attempt, but return to the camp at once."

During all this time of harrowing uncertainty Harold had shown remarkable self-control. After the first cry of horror, not a sound escaped his lips; but no one had dug into the snow with more desperate energy, and it was his hands that first touched his father's form. Now that the agony was over, the natural reaction followed, and his suppressed feelings found relief in a flood of tears as his father hugged him to his heart.

So rejoiced were all at the wonderful escape of Captain Marling and his companions, that there was

no room for regret at the mountain being left un-
scaled. Returning at once to the camp, they rested
there for the night, and the following day made their
way back to the *Narwhal* without further mishap.

There was still a month of waiting before them,
which they found very hard to endure patiently ere
the glad cry of "Open water in sight" came down
from the crow's nest, and it was a fortnight after that
before the open water made its way into the *Narwhal's* snug harbour. Then the fervid Arctic summer
seemed to come with a rush. The ice broke up into
floes as the warm wind blew upon it, the snow vanished from the hillsides before the hot rays of the
sun, and the great gulf, that had so lately been sleeping beneath its icy coverlet, awoke into waves that
danced and sparkled merrily in the sunshine, as
though they were challenging the noble vessel, which
like them had been for months asleep, to arouse herself and prepare for action.

The challenge was promptly accepted. The *Narwhal* was ready. Nothing remained to be done.
The deck-house had been cleared away, the top-masts
sent up, the engines oiled and burnished, and so on a
beautiful morning in June, amid the joyful cheers of
the crew, the sturdy steamer moved smoothly out into

the broad waters of the gulf, and the homeward journey was begun.

Harold stood beside his father on the bridge, his face beaming with joy.

"You look as though you were glad to be getting back home, Hal," said the captain, laying his hand upon his shoulder.

"I am indeed, father. I think we've had a splendid time of it up here, but I'll be so glad to see dear old Halifax again," answered Harold.

"You wouldn't care to change places with one of those Esquimaux, then?" inquired the captain, with a smile.

"Not by a long chalk," responded Harold promptly. "One winter in the ice is all well enough, but to spend one's whole life here—ugh! it would soon be the death of me."

Favoured by fair weather, the *Narwhal* made good progress up the gulf to Prince Regent Inlet, thence passing through Lancaster Sound and out upon the vast expanse of Baffin Bay, where, getting into the sweep of the great Arctic current setting strongly southward, she sped past Davis Strait to the Atlantic Ocean. The passage was marked by little incident. The constant presence of floe and berg rendered a careful look-out from the crow's nest always necessary; but every danger of this kind was successfully avoided, and by the latter part of August the steamer, having halted at Nachvak to land Lane the interpreter, was gliding past the forbidding coast of Newfoundland at a rate that three days later brought her into Halifax harbour.

Great was the interest her return created, and Captain Marling found himself quite a lion in nautical circles because of his remarkable voyage. Harold, too, came in for an embarrassing amount of attention, his former playmates looking upon him as a hero of the first rank. As for Patsy, when, arrayed in brand-new clothes, and with a pocket full of money, he appeared like a vision in the midst of the squalid surroundings from which he had fled with such happy results, his people utterly failed to recognize him, and he had some difficulty in persuading them that this stout, strong, brown-faced lad who bore himself so well was the lean, shambling, unkempt creature that they had known as Patsy Kehoe.

When the results of the voyage came to be ascertained, Captain Marling had no cause to regret his venture. The well-stored tanks yielded a rich return, and after dividing a handsome sum among the ship's company, everybody coming in for a share, the balance remaining to his credit was so considerable that he was enabled to carry out his design of giving up the sea, and settling down to a life on dry land for the remainder of his days, finding congenial employment in the owning and management of ships.

As regards Harold, however, the result of the voyage was precisely the opposite. It determined *his* future by making it clear beyond all question that the sea was his vocation. Accompanied by Patsy, who was ever to him what old Peter Strum had been to his father, he circumnavigated the globe in the captain's vessels, and in due time rose by steady gradations, until at length he trod the quarter-deck the proud commander of his own ship.

But far and wide as he sailed, and many and marvellous as were his adventures, wherever asked what was the most momentous and interesting experience of his life, he always answered that it was the voyage he took in the good steamer *Narwhal* up among the Arctic ice-floes.

THE END.

Travel and Adventure.

Jack Hooper. His Adventures at Sea and in South Africa. By VERNEY LOVETT CAMERON, C.B., D.C.L., Commander Royal Navy; Author of "Across Africa," etc. With 23 Full-page Illustrations. Price 4s., or with gilt edges, 5s.

"Our author has the immense advantage over many writers of boys' stories that he describes what he has seen, and does not merely draw on his imagination and on books."—SCOTSMAN.

With Pack and Rifle in the Far South - West. Adventures in New Mexico, Arizona, and Central America. By ACHILLES DAUNT, Author of "Frank Redcliffe," etc. With 30 Illustrations. 4s., or with gilt edges, 5s.

A delightful book of travel and adventure, with much valuable information as to the geography and natural history of the wild American "Far West."

In Savage Africa; or, The Adventures of Frank Baldwin from the Gold Coast to Zanzibar. By VERNEY LOVETT CAMERON, C.B., D.C.L., Commander Royal Navy; Author of "Jack Hooper," etc. With 32 Illustrations. Crown 8vo, cloth extra, gilt edges. Price 4s., or with gilt edges, 5s.

Early English Voyagers; or, The Adventures and Discoveries of Drake, Cavendish, and Dampier. Numerous Illustrations. Price 4s., or with gilt edges, 5s.

The title of this work describes the contents. It is a handsome volume, which will be a valuable gift for young persons generally, and boys in particular. There are included many interesting illustrations and portraits of the three great voyagers.

Sandford and Merton. A Book for the Young. By THOMAS DAY. Illustrated. Post 8vo, cloth extra. Price 2s. 6d.

Our Sea-Coast Heroes; or, Tales of Wreck and of Rescue by the Lifeboat and Rocket. By ACHILLES DAUNT, Author of "Frank Redcliffe," etc. With numerous Illustrations. Price 2s. 6d.

Robinson Crusoe. The Life and Strange Adventures of Robinson Crusoe, of York, Mariner. Written by Himself. *Carefully Reprinted from the Original Edition.* With Memoir of De Foe, a Memoir of Alexander Selkirk, and other interesting additions. Illustrated with upwards of Seventy Engravings by KEELEY HALSWELLE. Crown 8vo, cloth ex. 3s.

An edition that every boy would be pleased to include in his library. It is handsomely bound, and the numerous illustrations assist greatly in the realization of this famous story.

The Swiss Family Robinson; or, Adventures of a Father and his Four Sons on a Desolate Island. Unabridged Translation. With 300 Illustrations. Price 3s.

A capital edition of this well-known work. As the title suggests, its character is somewhat similar to that of the famous "Robinson Crusoe." It combines, in a high degree, the two desirable qualities in a book,—instruction and amusement.

Gulliver's Travels into Several Remote Regions of the World. With Introduction and Explanatory Notes by the late Mr. ROBERT MACKENZIE, Author of "The 19th Century," "America," etc. With 20 Illustrations. Post 8vo, cloth extra. Price 3s.

"A very handsome edition, under the editorship of Mr. Robert Mackenzie, who has supplied for it a well-written introduction and explanatory notes.... We have also here the curious original maps and a number of modern illustrations of much merit. Altogether this is a most attractive re-appearance of a famous book."—GLASGOW HERALD.

T. NELSON AND SONS, LONDON, EDINBURGH, AND NEW YORK.